Nick Vandome

Android Tablets

in easy steps

Covers Android 4.2

In Easy Steps Limited supports The Forest Stewardship Council (FSC),
the leading international forest certification organisation. All our titles
that are printed on Greenpeace approved FSC certified paper carry the
FSC logo.

MIX
Paper from
responsible sources
FSC® C020837

004.165
AND

Printed and bound in the United Kingdom

ISBN 978-1-84078-589-0

Contents

1 Introducing Android Tablets

Tablet computers and the Android operating system are an ideal match for anyone who wants their computing as mobile and as flexible as possible. This chapter introduces the basics of Android tablets and shows how to set up your tablet so that you can quickly get up and running with it.

About Tablets

Tablet computers are the result of the desire for our computing devices to become smaller and more portable (from desktops, to laptops, to tablets) and the evolution of mobile operating systems, initially introduced for smartphones. The combination of the two has resulted in the birth and relentless march of the tablet: they are small, portable for almost any situation, customizable and powerful enough to perform most everyday computing functions, such as word processing, email, using the Web and communicating with social networking sites. In addition, the functionality of tablets can be expanded almost endlessly through the inclusion of apps: programs that either come pre-installed or can be downloaded from a linked service.

Android Operating System

All computers need an operating system to make them work and perform all of the required tasks for the user. For tablets, the two main operating systems are iOS for Apple tablets (the iPad and iPad Mini) and Android. The latter has been developed through its use on smartphones and is now a significant player in the tablet market. The fact that it is used on both types of devices means that if you have an Android smartphone then an Android tablet is a perfect match.

Android is an open source operating system which means that developers and manufacturers can work with the source code to tailor it to their own needs and devices (as long as they meet certain requirements and standards). Android tablets are made by a number of different manufacturers and, although the hardware differs between devices, the Android operating system is common between them (although the versions of Android differ between devices, see page 10). Android tablets generally come in 7-inch or 10-inch models.

Touchscreen

Tablets are touchscreen devices, which means that their functionality and controls are accessed by tapping or pressing on the screen. This includes the keyboard, which appears on the screen if data input is required, e.g. for writing an email, entering a website address or filling in an online form. For people who have always used a physical keyboard, the virtual one can take a bit of getting used to, particularly if you are doing a lot of typing, but the more you use it, the more familar it will become.

Don't forget

Windows 8 can also be used on tablets and Microsoft have developed their own tablet, the Surface, which uses Windows 8.

Don't forget

Android is now owned by Google.

Hybrid tablets

In computing terms, the current tablet market is a relatively new one and manufacturers are developing ideas in terms of the evolution of the tablet, particularly in relation to the more traditional laptop. Some manufacturers have developed hybrid tablets that are designed to bridge the gap between tablets and laptops. These devices can be used as self-contained tablets, using the virtual, touchscreen, keyboard, or they can be docked to a physical keyboard and be used more like a laptop.

Don't forget

For more information about specific makes and models of Android tablets see pages 12–15.

As the tablet market develops and evolves further it is likely that there will be more hybrid models and, in time, tablets could begin to replace laptops for a lot of mobile computing needs.

Getting connected

One of the essential functions of tablets is online connectivity, for accessing the Web and also the range of Android services that are connected to Google apps and services.

The standard form of online connectivity for tablets is provided by Wi-Fi. This will connect to the Internet via your own Wi-Fi router and service provider in your home, or through a Wi-Fi hotspot if you are traveling with your tablet. Some models of tablets also have 3G/4G connectivity. This is wireless, mobile access to the Internet, provided by telecoms companies through either a monthly plan or pay-as-you-go options. This provides access to the Internet without the need for using Wi-Fi.

Don't forget

If your tablet has 3G/4G connectivity, you will have to pay for this service from an appropriate provider, in the same way as obtaining Internet access for a smartphone.

About Android

Android is essentially a mobile computing operating system, i.e. one for mobile devices such as smartphones and tablets.

Android is an open source operating system, which means that the source code is made available to hardware manufacturers and developers so that they can design their devices and apps in conjunction with Android. This has created a large community of Android developers and also means that Android is not tied to one specific device: individual manufacturers can use it (as long as they meet certain specific criteria) which leads to Android being available on a variety of different devices.

Android Inc. was founded in 2003 and the eponymous operating system was initially developed for smartphones. Google quickly saw this as an opportunity to enter the mobile phone and computing market and bought Android in 2005. The first Android powered smartphone appeared in 2008 and since then has gone from strength to strength. Android-based smartphones have a majority of the worldwide market and, with the increasing popularity of tablets, it is likely that those running Android will soon experience a similar level of success.

Updating Android

Since Android is open source and can be used on a variety of different devices, this can sometimes cause delays in updating the operating system on the full range of Android devices. This is because it has to be tailored specifically for each different device: it is not a case of 'one size fits all'. This can lead to delays in the latest version being rolled-out to all compatible devices. The product cycle for new versions is usually six to nine months.

Since Android is a Google product, their own devices usually are the first ones to run the latest version of the software. Therefore, the Google Nexus has been the first tablet to come with the latest version of Android, e.g. 4.2 Jelly Bean, while others are still running previous versions, usually 4.1 Jelly Bean or earlier. For recently-released tablets, an upgrade to the latest version of Android will be scheduled into the update calendar. However, for some older Android tablets, and smartphones, the latest version of the software is not always made available. This can be because of hardware limitations but there have also been suggestions that it is a move by hardware manufacturers designed to ensure that consumers upgrade to the latest products.

Don't forget

Android is based on the Linux operating system and shares many similarities with it.

Hot tip

Updated versions of Android are named alphabetically after items of confectionery, e.g. 1.5 was Cupcake and 1.6 was Donut; 4.0 was Ice Cream Sandwich and the latest versions, 4.1 and 4.2 are Jelly Bean.

Android apps

One of the great selling points for Android tablets is the range of apps that is available from third-party developers. Because Android is open source it is relatively easy for these developers to write apps for Android devices. At the time of writing there are approximately 700,000 Android apps on the market. Some are free while others have to be paid for.

The pre-installed apps are the ones that give the initial functionality to your tablet and include items such as email, web browser, calendar, notes and maps. They appear as icons on your tablet's Home screen, or in the All Apps area, and are accessed by tapping lightly on them once.

Don't forget

The Kindle Fire tablet runs on what is known as a 'forked' version of Android. This is based on an ealier version of Android that was then developed separately. For a detailed look at the Kindle Fire, see Get Going with Kindle Fire In easy steps.

New apps for Android tablets are available through the Play Store, or directly from the developer's website. They can be downloaded from here and then appear on your tablet.

Unless specified otherwise, Android apps are self-contained and do not interact with each other on your tablet. This has reduced the risk of viruses spreading through your tablet and also contributes to its memory management. When you switch from one app to another you do not have to close down the original one that you were using. Android keeps it running in the background, but in a state of hibernation so that it is not using up any memory or processing power on your tablet. However, when you want to use it again it will open up at the point at which you left it. If your tablet is running low on memory it will automatically close any open apps to free up more memory. The ones that have been inactive for the longest period of time are the ones that are closed first, until enough memory has been freed up.

Makes and Models

Due to the open source nature of Android it can be used on a range of different devices and several manufacturers use it on their tablets. Two of the main Android tablets on the market are the Google Nexus and the Samsung Galaxy Tab 2, which both have 7- and 10-inch versions.

Google Nexus (7- and 10-inch)

Google Nexus. Google's flagship tablet, running the very latest version of Android, comes in 7-inch and 10-inch versions. The hardware is produced by Asus but it is very much a Google product. The main features are:

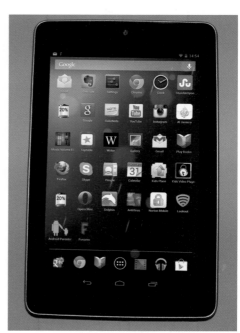

- Screen: 7", 1280 x 800 HD display (216 PPI) back-lit IPS; 10", 2560 x 1600 (300 PPI)

- Dimensions: 198.5 x 120 x 10.45 mm (7"); 263.9 x 177.6 x 8.9 mm (10")

- Weight 340g (7"); 603g (10")

- Processor: NVIDIA® Tegra® 3 quad-core processor with 1GB RAM (7"); Dual-core ARM Cortex-A15 processor with 2GB RAM (10")

- Storage: 16GB/32GB of internal storage

- Wireless: WiFi 802.11 b/g/n. Bluetooth NFC (Android Beam)

- Operating System: Android 4.2 (Jelly Bean)

- Battery power: Up to 8 hours of active use

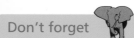

Don't forget

The main controls on the Google Nexus (Back button, Home screen button and Recently Used Apps button) appear at the bottom, middle of the screen. Tap on the screen if they are not visible.

- Battery charging: Via power adapter (supplied) or micro USB to computer system

- Camera: 1.9 MP (front) (7"); 5 MP (rear), 1.9 MP (front) (10"). Although the front facing cameras can be used for taking photos, they are generally used for making video-calls. If you want to take good quality photos on your tablet then it is best to use one with a rear-facing camera and a high megapixel (MP) number

- Input/Output: Micro USB cable: 3.5 mm headphone jack, built-in speaker and microphone

- Sensors: Accelerometer, GPS and gyroscope

Nexus Controls

As with most tablets, the physical controls (on/off, volume) are located around the side of the Nexus (see page 18 for details). There are also various touchscreen controls that are located at the bottom of the screen:

1 Frequently-used apps can be pinned in the Favorites Tray, at the bottom of the screen, above the navigation buttons

2 Use these buttons to, from left to right, go back one screen, go to the Home screen and view recently used apps

3 Use this button to view All Apps

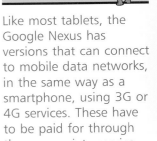

Don't forget

Like most tablets, the Google Nexus has versions that can connect to mobile data networks, in the same way as a smartphone, using 3G or 4G services. These have to be paid for through the appropriate service provider, and this will provide mobile Internet access even when you do not have Wi-Fi access.

Don't forget

On some models of tablet, the All Apps button is located in the top right-hand corner of the screen.

...cont'd

Samsung Galaxy Tab 2 (7- and 10-inch)

Samsung Tab. With their experience of using Android on their range of Galaxy smartphones, Samsung were well placed to enter the tablets market and they have done this successfully with their Tab range of 7-inch and 10-inch tablets.

The main features are similar to those of the Google Nexus but, at the time of printing, it runs on a slightly earlier version of Android, 4.1 Jelly Bean, or earlier.

Samsung Tab 2 Controls

The touchscreen controls on the Samsung Tab 2 are also similar to the Google Nexus, except that they are located in the bottom left-hand corner on the screen. The button are, from left to right, Back button, Home screen button, Recently-used Apps button and the Screenshot capture button.

14

Samsung Note

Although this is similar in some respects to the Tab, it has a more powerful processor and much of its functionality is done with an S Pen, or stylus. (See page 26.) This can be used to write and draw directly on the screen.

Asus Transformer

This is an example of manufacturers developing hybrid tablets, i.e. ones which can also be used as a more traditional laptop. In the case of the Transformer, this is done with a detachable keyboard/docking station.

Lenovo IdeaTab

This comes in 9-inch and 7-inch models and, in addition to apps from the Google Play Store, also provides access to apps in the Lenovo App Shop.

Don't forget

The Home screens of the tablets on this page may have their own apps included. However, all of the generic Android ones can be downloaded from the Google Play Store, which is accessed from the Play Store button.

Sony Tablet S

A well specified tablet that also comes in a stylish wedge-shaped design and provides access to Playstation games.

Toshiba AT300

A thin 10-inch tablet, this is a good performer in the Android tablet market.

Motorola Xoom 2

One of the first Android tablets, this version continues to provide good performance and features for the tablet user.

Android and Google

Most tablets are linked into a specific company for the provision of their services and selection of apps: Apple for the iPad, Amazon for the Kindle Fire and Google for tablets using Android. As with the other tablets, for Android ones you must have a linked account to get the most out of your tablet. This is a Google Account and is created with a Google email address (Gmail) and a password. Once it has been created it will give you access to a number of the built-in Android apps and also additional services such as backing up and storing your content.

When you first set up your Android tablet you can enter your Google Account details, or select to create a new account. You can also do this at any time by accessing one of the apps that requires access to a Google Account. These include:

- Play Store, for obtaining more apps

- Play Movies

- Play Books

- Play Magazine

- People, for an online address book. When you enter contact details these are made available from any web-enabled device

Other apps, such as the Gallery for storing and viewing photos, can be used on their own, but if a Google Account has been set up their content will be backed up automatically.

Some of the benefits of a Google Account include:

- Access from any computer or mobile device with web-access, from the page **account.google.com/**
  Once you have entered your account details you can access the **Products** section, including your Calendar, Gmail and Play Store

- Keep your content synchronized and backed up. With a Google Account, all of your linked data will be automatically synchronized so that it is available for all web-enabled devices and it will also be backed up by the Google servers

- Peace of mind that your content is protected. There is a **Security** section on your Google Account web page where you can apply various security settings and alerts

Creating a Google Account

A new Google Account can be created in the following ways:

- During the initial setup of your Android tablet

- When you first access one of the relevant apps, as shown on the previous page

- From the **Settings > Accounts** option

For each of the above, the process for creating the Google Account is the same:

1 On the **Add a Google Account** screen, tap on the **New** button

2 Enter the first and last name for the new account user and tap on the right-pointing arrow

3 Enter a username (this will also become your Gmail address) and tap on the right-pointing arrow

4 Create a password for the account and then re-enter it for confirmation

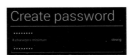

5 On the **Finish account** page, check on the required items and tap on the right-pointing arrow to access the authentication screen

Buttons and Ports

Although the buttons and ports on Android tablets will differ slightly between devices, the basic functionality is similar.

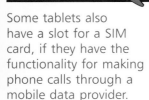
On/Off button. This can also be used to put the tablet into Sleep mode. Press and hold for a couple of seconds to turn on the tablet. Press once to put it to sleep or wake it up from sleep

Volume button. Press at the ends to increase or decrease volume

Headphone jack   **Speaker**

Micro USB port (this can be used to attached the tablet to an adapter for charging, or to a computer for charging or to download content from the tablet, using the supplied USB cable)

Battery and Charging

All tablets run on internal batteries which usually offer up to approximately 10 hours of average usage.

Tablet batteries can be charged with a USB adapter that connects via the tablet's micro USB port, with a supplied cable. This can also be connected to a desktop or laptop computer, but this takes longer to charge the tablet than using the dedicated adapter.

To charge a tablet's battery:

1 Attach the micro USB cable to the tablet

2 Connect the USB cable to the adapter, using the USB jack

3 Connect the adapter to the mains power. It is best to charge it fully before disconnecting it from the mains power, although it can still be used while it is being charged

Setting Up Your Tablet

When you first turn on your tablet (by pressing and holding the On/Off button) you will be taken through the setup process. This only has to be done once and some of the steps can be completed, or amended, at a later time, usually within the **Settings** app. The elements that can be applied during the setup process are:

- **Language**. This option lets you select the language to use for your tablet. Whichever language is selected will affect all of the system text on the tablet and it will also apply to all user accounts on the tablet

- **Wi-Fi**. This can be used to set up your Wi-Fi so that you can access the Web and online services. In the **Select Wi-Fi** window, tap on the name of your router

Enter the password for your router and tap on the **Connect** button

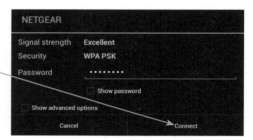

Don't forget

Most routers require a password when they are accessed for the first time by a new device. This is a security measure to ensure that other people cannot gain unauthorized access to your router and Wi-Fi.

- **Google Account**. At this stage you can create a Google Account, or sign in with an existing one. Once you have done this you will have full access to the Google Account services and you will not have to enter your login details again

- **Google+**. This is a service that can be used to share content, such as photos, with other users of Google+. Tap on the **Join Google+** button to join this service

- **Backup and Restore.** In this section you can specify options for how the content on your tablet is backed up, and how it can be restored if it is lost. This includes specifying that your apps data and other settings are backed up to the Google servers and can also be automatically restored

- **Google & Location**. This provides options for selecting whether to let apps use your location for collecting data and enhancing the functionality of the apps. This information is sent to Google. Tap on the boxes next to the relevant options to activate each one

- **Setup complete**. Once the setup has been completed tap on the arrow to move to the Home screen of your tablet

Don't forget

For details about the items on the Home screen, and navigating around it, see Chapter Two.

Adding Multiple Users

With some versions of Android (4.2 and later) it is possible to add multiple users on a tablet. This is a very useful function as it allows different people to have their own private space on the tablet where they can access their own content and apply the settings that they want.

The person who first sets up the tablet is the owner of it and they have ultimate control in terms of adding and deleting other users. To add a new user:

Hot tip

For more information about accessing and working with Android apps, see Chapter Two.

1 Tap on the **All Apps** button

2 Tap on the **Settings** app

Settings

Hot tip

For Android 4.2 and later, the Settings app can also be accessed by swiping down from the top right-hand corner of your tablet and tapping on the **Settings** button. For more information on accessing and using Notifications and Settings on your tablet, see Chapter Two.

3 Under the **Device** section, tap on the **Users** button

👤 **Users**

4 Your information is displayed as the current user, and owner, of the tablet

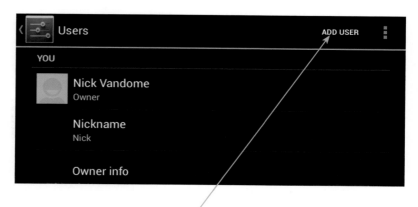

5 Tap on the **Add User** button

6 Tap on the **OK** button

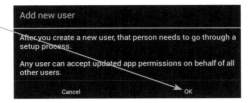

7 Tap on the **Set up now** button (the new user will now have to enter their details)

8 Once the new user has added their details the **Lock** screen will include a button for the new user. If it is magnified it indicates that this is the Lock screen for that user

9 Drag the padlock icon to the edge of the circle to unlock the screen for the new user

10 The **Welcome** screen notifies you that you have been added as a new user on the tablet. Tap on the arrow to move through the setup process. This is similar as for creating a user when the tablet is first set-up as on pages 20-21

Hot tip

Each individual user can set a password that has to be entered to unlock the tablet for access to their own content. For more details about this, see Chapter Two.

Accessibility

It is important for tablets to be accessible for as wide a range of users as possible, including those with vision or physical and motor issues. In Android this is done through the **Accessibility** settings. To use these:

1 Tap on the **Settings** app

2 Under the **System** section, tap on the **Accessibility** button

3 Tap on an item to turn it On or Off, or check on these boxes to enable functionality for increasing the text size, auto-rotating the screen and having passwords spoken as they are entered

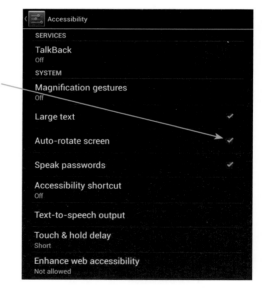

4 Tap on the **TalkBack** button in Step 3 and drag this button to **On** to activate TalkBack, whereby the tablet will provide spoken information about items on screen and those which are being accessed

5 TalkBack also provides an **Explore by Touch** function that enhances TalkBack by providing an

audio description of what is on the screen. Tap on the **OK** button to activate this

6 The items on the screen are described by audio. Tap on an item to hear an audio description.

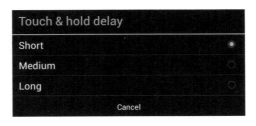

Double-tap to activate a feature

7 Tap on the **Touch & hold delay** button in Step 3 to select the time delay for an action to take

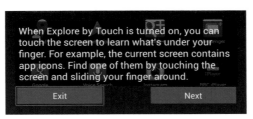

effect when you press and hold on the screen or an icon

8 Tap on the **Enhance web accessibility** button in Step 3 and tap on **Allow** to allow apps to install scripts from Google to make their web content more accessible. This could include allowing their text size to be increased and enabling screen reader functionality so that the content can be spoken

Do you want apps to install scripts from Google that make their web content more accessible?

Don't allow   Allow

Adding Accessories

As with most electrical gadgets, there is a wide range of accessories that can be used with Android tablets. Some of these are more cosmetic, while others provide useful additional functionality. Some to consider are:

- **Docking station.** This can be used to attach to your tablet and it doubles as a stand for viewing content and also for charging your tablet

- **Stylus pen.** This is a pen with a rubber tip that can be used to write on a tablet, tap on items to activate them and also swipe between screens or pages

- **Battery charging pack.** This is a mobile unit that can be used to charge your tablet when you are away from a mains source of power. The pack is charged initially (indicator lights show how much charge is available) and it can then be plugged into the tablet to give it additional power

- **Cover.** This can be used to protect the tablet and, in some cases, they double as a stand for viewing content on the tablet

- **Screen protector.** If you want to give your tablet's screen extra protection, these sheets of clear plastic are a good option. Some of them also come with cleaning cloths

- **USB adapter.** This is an adapter that connects to your tablet's micro USB connector so that USB devices can be connected to it. This can include digital cameras, pen drives and card readers for photos

- **Mobile Wi-Fi unit.** This is a unit that can provide Wi-Fi access when you are away from your own Wi-Fi router. They are used with pay-as-you-go SIM cards so you only pay for what you use and do not need to have a long-term contract

26

2 Around an Android Tablet

The interface for an Android tablet is more similar to that of a smartphone than a traditional computer. However, it has many similar features to a computer and can perform a lot of the comparable tasks. This chapter details the Android interface for tablets and shows how to find your way around the Home screen, add apps and widgets, access your notifications and lock your tablet for peace of mind.

Viewing the Home screen

Once you have set up your tablet the first screen that you see will be the Home screen. This is also where you will return to when you tap the Home button (see next page). The elements of the Home screen are:

For some versions of Android (Jelly Bean 4.2 and later) there is also a **Quick Settings** area that can be accessed by swiping down from the top right-hand corner of the screen. See page 37 for more details.

Don't forget

On some models of tablet, the All Apps button is located in the top right-hand corner of the screen.

Notifications area

Google Search box

Home screen area. This is where the majority of your commonly-used apps and widgets will be located

Favorites Tray

Navigation buttons

All Apps button

On different models of Android tablet, the Navigation buttons and All Apps button may be located in a slightly different position. The appearance of the Home screen may also be slightly different, depending on which apps the manufacturer wants to appear on the Home screen.

28

Navigating Around

At the bottom of the Home screen are three buttons that can be used to navigate around your tablet. These appear on all subsequent pages that you visit so that you can always use them for navigation purposes.

The Navigation buttons are:

Back. Tap on this button to go back to the most recently-visited page or screen

Home. Tap on this button to go back to the most recently-viewed Home screen at any point

Recent apps. Tap on this to view the apps that you have used most recently. Tap on one of the apps to access it again

Don't forget

Most Android tablets have several Home screens. Swipe left and right to move between them.

When the keyboard is being used the Back button turns into a down-pointing arrow. Tap on this once to hide the keyboard and reveal the Back button again.

Going back

As well as using the Back button to return to the previous screen this can also be done, in some cases, via the icon for the relevant app at the top of the screen. For instance, if you are in the **Settings** app you can return to the previous screen by tapping on this icon.

Editing My Library

One of the first items that is visible on the Home screen is the My Library widget. This displays the most recent items that you have downloaded from the online Google Play Store or, if you have not yet downloaded anything, it will contain recommendations for you. The My Library widget can be resized on the Home screen and also removed if required. To do this:

Hot tip

Tap once on an item within the **My Library** widget to open it.

1 Press and hold within the **My Library** widget to activate the blue border

Don't forget

The **My Library** widget is located within the **Widgets** area within the **All Apps** section.

2 Drag the border to resize the **My Library** widget

3 To delete the **My Library** widget, press and hold on it until the **Remove** button appears at the top of the screen

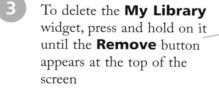

Don't forget

The widget is only removed from the Home screen not the tablet itself.

4 Drag the widget over the **Remove** button until it turns red. Release to delete the widget

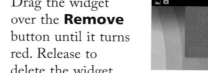

Adding Apps

The Home screen is where you can add and manage your apps.
To do this:

1 Tap on the **All Apps** button

2 All of the
built-in apps are
displayed. Tap
on an app to
open it

Don't forget

There are thousands
more apps available for
download from the Play
Store. Tap on this button
in the Favorites Tray to
access the Play Store.

3 To add an app
to the Home
screen, tap and
hold on it

4 Drag it onto
the screen on
which you want
it to appear and
release it

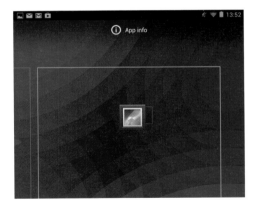

5 The app is added
to the Home
screen

6 Swipe left and
right to move
between the
available Home
screens

Don't forget

Apps can be removed
from a Home screen
in the same way as
removing the My Library
widget as shown on the
previous page.

Moving Apps

Once apps have been added to the Home screen they can be
repositioned and also moved to other Home screens. To do this:

Apps can be moved to
the left or right onto
new Home screens.

Beware

Make sure that the app
is fully at the edge of the
Home screen, otherwise
it will not move to the
next one. A thin blue
border should appear
just before it moves to
the next Home screen.

1 Press and
hold on an
app to move
it. Drag it
into its new
position. A
blue outline
will appear
indicating
where the app will be positioned

2 Release the
app to drop
it into its
new position

3 To move an
app between
Home
screens,
drag it to
the edge of
the Home
screen

4 As the app reaches the edge
of the Home screen it will
automatically move to the next
one. Add it to the new Home
screen in the same way as
in Step 2

Working with Favorites

The Favorites Tray at the bottom of the Home screen can be used to access the apps you use most frequently. This appears on all of the Home screens. Apps can be added to, or removed from, the Favorites Tray, as required.

1 Press and hold on an app in the **Favorites Tray** and drag it onto the Home screen. A gap appears where the app has been removed

Don't forget

For some tablets, the **Favorites Tray** appears along the bottom of the screen in landscape mode; for others it appears down the right-hand side.

2 Tap and hold on the apps in the **Favorites Tray** and drag them into new positions as required

Hot tip

Apps can appear in the **Favorites Tray** and also on individual Home screens, but they have to be added there each time from within the **All Apps** section.

3 Press and hold on an app and drag it onto a space in the **Favorites Tray** to add it there

4 The **Favorites Tray** has a limit to the number of apps that it can contain (usually six) and if you try to add more than this the app will spring back to its original location

Adding Widgets

Android widgets are similar to apps, except that they generally display specific content or real-time information. For instance, a photo gallery widget can be used to display photos directly on a Home screen, or a traffic widget can display updated information about traveling conditions. Widgets can be added from within the All Apps section:

1 Tap on the **All Apps** button

2 Tap on the **Widgets** tab

Don't forget

Widgets on a Homes creen are usually larger than apps but they can be resized as shown with the **My Library** widget on page 30.

3 Swipe left and right to view all of the available widgets

4 Press and hold on a widget and drop it onto a Home screen as required

Changing the Background

The background (wallpaper) for all of the Home screens on your tablet can be changed within the **Settings** app (**Settings > Wallpaper**). However, it can also be changed directly from any Home screen. To do this:

1 Press and hold on an empty area on any Home screen

2 Tap on one of the options from where you would like to select the background wallpaper

3 Swipe left and right to select a background and tap on the **Set wallpaper** button

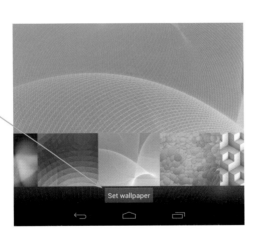

The **Live Wallpaper** option contains animated backgrounds rather than just static ones.

35

4 The selected background is applied to all Home screens

Notifications Bar

Android tablets have numerous ways of keeping you informed about what is happening on your device; from new emails and calendar events, to the latest information about apps that have been downloaded and installed. To make it easier to view these items, they are grouped together on the Notifications Bar, from where you can then access them directly. This is located at the top left-hand corner of the screen:

Don't forget

On some tablets, the Notifications Bar is located at the bottom right of the screen and it is accessed by swiping up from the bottom of the screen.

Don't forget

If you clear the notifications it does not delete the items; they remain within their relevant app and can be viewed there.

 Current notifications are identified by specific icons on the **Notifications Bar**

 Swipe down to view details about the current notifications

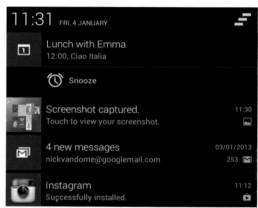

Tap here to clear all current notifications

Tap on a notification to view its full details, within the relevant app

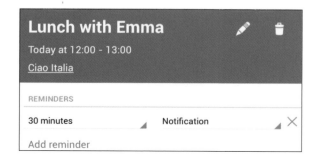

Quick Settings

The full range of Android settings can be accessed from the Settings app (see Chapter Three for full details). However, for Android 4.2, and later, there is a Quick Settings options that can be accessed from the top of the screen. To use this:

1 Swipe down from the top right-hand corner of the screen (opposite the Notifications Bar)

2 All of the **Quick Settings** options are displayed

Hot tip

The Quick Settings, and Notifications Bar, can be accessed if the tablet is locked, but only with the Slide method (see page 40 for details).

3 Tap on a setting to access its options in the **Settings** app, or

4 Tap on an item to apply it directly within

Quick Settings. This can be done with **Aeroplane Mode**, **Auto Rotate** and **Brightness**. The **Brightness** option activates a separate widget that can be used to set the screen brightness

Creating Folders

As you start to use your Android tablet for more activities, you will probably acquire more and more apps. These will generally be for a range of tasks covering areas such as productivity, communications, music, photos, business and so on. Initially it may be easy to manage and access these apps, but as the number of them increases it may become harder to keep track of them all.

One way in which you can manage your apps is to create different folders for apps covering similar subjects, e.g. one for productivity apps, one for entertainment apps etc. To do this (Android 4.2 and later):

38

1 Press and hold on an app and drag it over another one

2 Release the app. A folder will be created, containing both of the apps

3 Tap on a folder to view its contents. Initially it will be unnamed

4 Tap here and give the folder a relevant name

5 Folders can contain numerous items and also be placed in the **Favorites Tray**

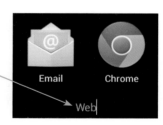

Screen Rotation

By default, the content on a tablet's screen rotates as you rotate the device. This means that the content can be viewed in portrait or landscape mode, depending on what is being used, e.g. for movies it may be preferable to have it in landscape mode, while for reading it may be better in portrait mode:

Don't forget

Screen rotation is achieved by a gyroscope sensor in the tablet.

It is also possible to lock the screen so that it does not rotate when you move it. This can be useful if you are using it for a specific task and do not want to be distracted by the screen rotating if you move your hand slightly. To lock and unlock the screen rotation:

1 Drag down from the top right-hand corner of the screen to access the **Quick Settings** (on some versions of Android this will be the general settings that are available by dragging down from the top of the screen)

2 Tap on the **Auto Rotate** button to lock screen rotation

3 Tap on the **Rotation Locked** button to disable the screen lock and return to Auto Rotate mode

Don't forget

For some tablets, the screen rotation is accessed by swiping up from the bottom of the screen and tapping on the Screen Rotation button.

Locking Your Tablet

Security is an important issue for any computing device and this applies to physical security as much as online security. For Android tablets it is possible to place a digital lock on the screen so that only an authorized user can open it. This is particularly useful when there are multiple user accounts on the same tablet. There are different ways in which a lock can be set.

1 Tap on the **Settings** app

2 In the **Personal** section, tap on the **Security** button

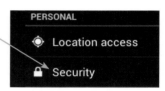

3 The current method of **Screen lock** is displayed here. Tap on this to access the options

4 The methods for locking the screen are **Slide**, **Face Unlock**, **Pattern**, **PIN** and **Password**. Tap on the required method to select it and set its attributes

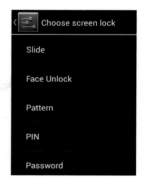

Beware

The Slide option is only really useful for avoiding items being activated accidentally when your tablet is not in use; it is not a valid security method. The most secure method is a password containing letters, numbers and characters.

5 The **Slide** option is the least secure as it only requires the padlock icon to be slid to the edge of the circle to unlock the tablet. No other security authorization is required

6 For the **PIN** (or **Password**) option, enter the PIN in the box and tap on the **Continue** button. Enter the PIN again for confirmation. This will then be required to be entered whenever you want to unlock the tablet

7 For the **Pattern** option, drag over the keypad to create the desired pattern

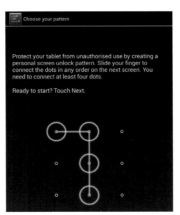

41

8 For the **Face Unlock** option, look at the camera on the tablet. A green dotted outline will be created around your face. When you want to unlock your tablet, look at the required area on the screen. If your face is not recognized, a PIN or Password can be used as an alternative

Lock Screen Apps

When the screen of your tablet is locked it is still possible to access certain apps and types of content. In Slide lock mode the Notifications Bar and Quick Settings can be accessed by swiping down from the top of the screen and it is also possible to add some apps to the Lock screen in all lock modes. To do this:

Don't forget

Lock screen apps are available with Android 4.2 and later.

Hot tip

Each user can also enter text that appears on the Lock screen. This is done in **Settings > Users > Owner** (or **User**) **Info**.

Beware

If your tablet is locked by any method other than Slide, the tablet will not show up as an external drive if it is connected to a computer, until it is unlocked.

1 On the Lock screen tap on the **+** sign (if this is not showing, swipe left or right to display it)

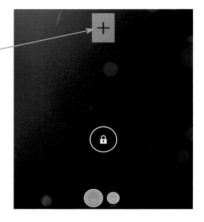

2 Tap on an app to add it to the Lock screen

3 The app is displayed on the Lock screen, regardless of the method of screen lock that has been applied. However, the screen has to be unlocked before the app can be used, other than for viewing the information on it

Searching

Since Android is owned by Google, it is unsurprising that tablets with this operating system come with the power of Google's search functionality. Items can be searched for on the tablet itself, or on the Web. This can be done by typing in the Google Search box and also by using the voice search option. To search for items on an Android tablet:

1 By default, in portrait mode, the Google Search box appears at the top of every Home screen

2 In landscape mode, the search option is minimized at the left-hand side. Tap on the **g** symbol to activate the Search box

3 The **Google** app can also be used for searching. Access this in the **All Apps** section and tap on it once to activate the Google Search box

4 Begin typing a word or phrase. As you type, corresponding suggestions will appear, both for on the Web and for apps on the tablet

5 As you continue to type the suggestions will become more defined

Hot tip

The Google Search box can also be accessed by swiping up from the bottom of the screen, from any app. This also displays the Google Now service, if it has been activated on your tablet.

Don't forget

On some tablets, the search option is indicated by a magnifying glass with the word Google next to it.

...cont'd

6   Tap on an app result to open it directly on your tablet, or tap on this button on the keyboard to view the results from the web

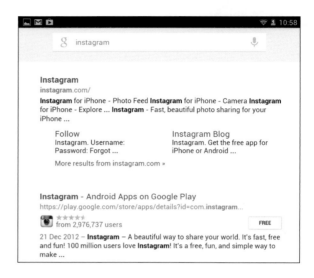

Voice search

To use the voice search functionality on your tablet, instead of typing a search query:

1   Tap on the microphone button in the Search box

2   When the microphone button turns red, speak the word or phrase for which you want to search. The results will be displayed in the same way as for a typed request

Investigating Settings

This chapter looks at the Settings that can be applied on an Android tablet so that you can customize your device the way you want.

Accessing Settings

We all like to think of ourselves as individuals and this extends to the appearance and operation of our electronic gadgets. An Android tablet offers a range of settings so that you can set it up exactly the way that you want and give it your own look and feel. These are available from the **Settings** app and cover settings for:

- Wireless & Networks
- Device
- Personal
- Accounts
- System

To access the **Settings** on your Android tablet:

1. Tap on the **All Apps** button

2. Tap on the **Settings** app

3. The full range of settings is displayed

4. Tap on an item to view all of the options for it (if necessary, tap on the options at the next level down to see their own options). Most options will have an On/Off button, a radio button or a checkbox to tap on or off

Wireless and Networks

Wi-Fi

Wi-Fi settings can be used to connect to your Wi-Fi router, for online access, and also apply advanced settings.

1   Slide this button **On** to enable Wi-Fi on your tablet

2   Tap here to connect your router and set up a wireless network

3   Tap on the name of your Wi-Fi router to connect it

4   Enter the password for the router and tap on the **Connect** button

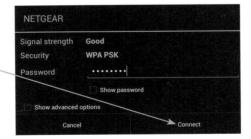

5   Tap on these buttons at the top of the Wi-Fi window to, from left to right, refresh the current router, add another Wi-Fi network and access additional settings

Don't forget

If you are accessing Wi-Fi away from your home router via a Wi-Fi hotspot, this will show up as a new router in Step 3. In some cases you will require the password from the provider of the Wi-Fi hotspot. For others, the connection may be made automatically.

...cont'd

Bluetooth

The Bluetooth settings can be used to connect your tablet to other Bluetooth devices, such as smartphones. To do this:

1 Slide this button **On** to enable Bluetooth on your tablet

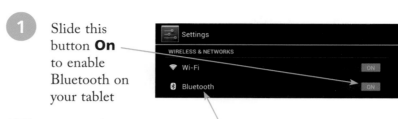

Beware

To pair your tablet with another device, the other device has to have Bluetooth enabled and be made visible for pairing for other devices.

2 Tap here to setup Bluetooth and connect your tablet to another device. This is known as pairing

3 Tap here to make your tablet visible to the other devices

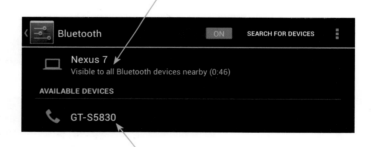

4 Any available devices are displayed here

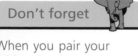

Don't forget

When you pair your tablet via Bluetooth the other device should display the passkey displayed in Step 5. This is generated each time that a device is paired.

5 Tap on the name of another device to pair your tablet with it. Tap on the **Pair** button

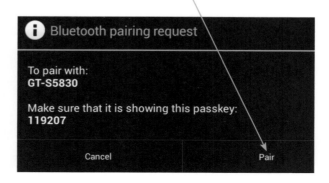

6 Paired devices are shown here

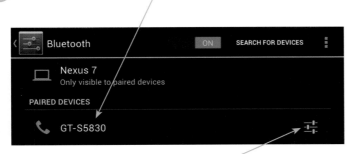

7 Tap here to access the settings for a paired device. This can then be used to **Unpair** the device, if required

> **Don't forget**
>
> When two devices have been paired, Bluetooth can be used to share items such as photos and music, wirelessly, over short distances, approximately 20 metres.

Sharing with Bluetooth

Once your tablet has been paired with another Bluetooth device you can then share items such as photos. To do this:

1 Select an appropriate item, such as a photo in the **Gallery** app. Tap on the **Share** button, or the **Bluetooth** button if it is available

2 Tap on the device with which you want to share the item. On the paired device you will have to **Accept** the request to share an item via Bluetooth

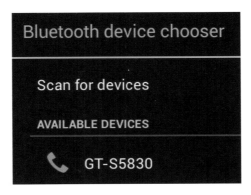

> **Don't forget**
>
> If you turn off Bluetooth on your tablet, paired devices should still be visible when you turn it back on, as long as you have not unpaired a device as in Step 7.

49

...cont'd

Data usage

The **Data usage** settings can be used to show how much data is being used by individual apps. This is particularly useful if you have an Android that has 3G or 4G capability and has a limit on the amount of data that can be downloaded in this way. To use the Data usage settings:

1 Tap on the **Data usage** button

2 The overall usage for a time period is shown at the top of the window. Tap here to select another time period to display

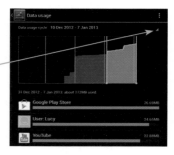

3 Tap on an app to view details about its usage. Drag these arrows to show the usage for a greater, or lesser, time period

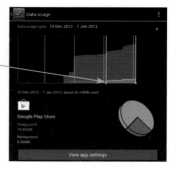

 Don't forget

The button in Step 4 can also be used for accessing the options for all of the other **Settings** too.

4 Tap on this button to access the menu options for the Data usage

More settings

Under the Wireless & Network settings there is also a **More** option, for additional settings, usually depending on the make and model of tablet:

1 Tap on the **More** button to access these settings

2 Some of the options may include **Aeroplane mode**, which can be checked on so that your tablet does not transmit data when you are flying, and **VPN** which can be used to set up a Virtual Private Network (usually for business users)

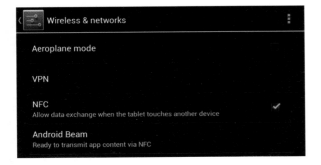

3 Check on the **NFC** option and tap on the **Android Beam** button to turn on this option. This can be used to transfer items between compatible devices

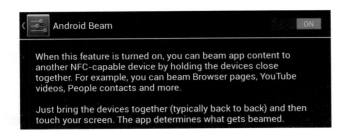

Don't forget

Android Beam can be used to transfer files between devices by touching them back-to-back when an appropriate app is open.

Device

The Device settings can be used to change the settings on your tablet for system sounds, display brightness, storage, battery usage, apps and individual users.

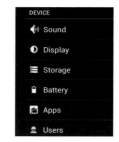

Sound

These settings can be used to set the volume for different items, specify a notification sound and turn on, and off, system sounds, such as for when the screen, or lock screen, is touched.

Hot tip

Turning off some of the system sounds may be one of the first things that you do on your tablet if you do not like hearing sounds each time you touch the keyboard or screen.

1 Tap on the **Sound** button under **Device**

2 Tap here to enable, or disable, the sound for when the screen is touched

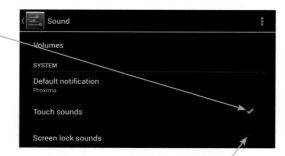

3 Tap here to enable, or disable, the sound for when the lock screen is unlocked or put to sleep

4 Tap on the **Volumes** button and drag the buttons to set the volume for the relevant items

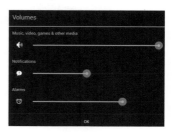

5 Tap on the **Default notification** button to select a sound for when notifications arrive

Display
To set different options for your tablet's display:

1 Tap on the **Display** button under **Device**

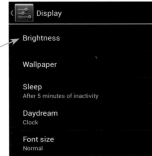

2 Tap on the **Brightness** button and drag the button to change the screen brightness. Tap on the **OK** button

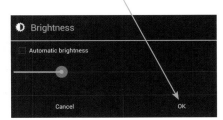

53

3 Tap on the **Sleep** (timeout) button and select a time period for the length of inactivity until the tablet goes into sleep mode

4 Tap on the **Daydream** button and tap on an item for this function. This appears when the tablet is charging or docked

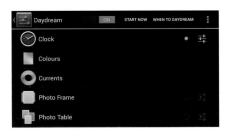

...cont'd

Storage

This displays the amount of space that is being taken up by different types of content on your tablet.

1 Tap on the **Storage** button under **Device**

2 The total amount of storage used is shown at the top of the window. The amount used by different types of content is shown here. Tap on an item to access it directly

---

Don't forget

Items such as apps, music, videos and photos tend to take up most of the space on your tablet.

Battery

This shows how much power is used up by certain functions and apps. To view these details:

1 Tap on the **Battery** button under **Device** to view the different functions and apps that are using battery power

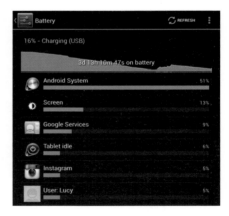

2 Tap on an item to see additional details for it. For some items, such as the screen, the power use settings can be edited, i.e. tap on the **Display** button to adjust the screen brightness

Apps (or Application Manager)

This can be used to view details about all of the apps on your
tablet. This includes the system apps, that are pre-installed, and
those that you have downloaded:

1   Tap on the **Apps** button under **Device**

2   Tap on the tabs at the top
of the window to view
apps or **Downloaded**,
Running and **All**

3   Tap on an app to view
additional details about it,
such as the storage taken
up by the app and the
data is has stored. You can
also **Force stop** the app
and clear its data and
stored cache

Beware

Some apps, such as the
Maps and Navigation
ones, keep accessing
data in the background
if they have been opened
but are not currently
being used. To conserve
battery power, use the
Force stop option to
close these apps when
not in use.

Users (only available with Android 4.2 or later)

This setting can be used to view, and edit, details about all of the
users on your tablet.

1   Tap on the **Users** button under **Device**

2   Tap on a user's name
to view their details.
Tap on the **Owner
info** button to add text
that is visible on the
Lock screen

Personal

The Personal settings can be used to specify how the information on your tablet is shared, security settings, language and input options and settings for how your content is backed up.

Location access (or Location Services)

 Tap on the **Location access** button under **Personal**

 Drag this button **On** to give your apps permission to use your location information. In some cases this will enable apps to target you with location-specific information

3 Tap on these buttons to give permission for apps to detail your location using GPS and Wi-Fi

Security (for Android 4.2 or later)

1 Tap on the **Security** button under

2 Under **Screen Security**, tap on **Screen lock** and **Owner info** to select options for these

3 Tap on the **Device administrators** button to see who can perform administration tasks on the tablet. This will usually be done if the tablet is being used in a corporate environment, such as using Exchange email

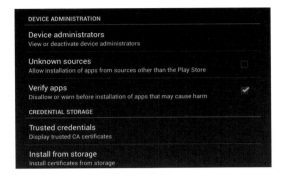

The **Verify apps** option can be used to check apps for any harmful content before they are installed. This is only available with Android 4.2 and higher.

4 Tap on the **Unknown sources** checkbox to allow apps to be installed from sources other than the Play Store

Language & input

1 Tap on the **Language & input** button under **Personal**

2 Tap on the **Language** button to select the language used on your tablet. Tap on the **Spell checker** checkbox to activate this for use with text input on the keyboard

3 Tap on the **Google voice typing** checkbox to enable text input to be done by speech rather than typing. When this is enabled a microphone appears on the keyboard

...cont'd

4 Tap on the **Settings** button for Google voice typing in Step 3 on page 57 and select options for this functionality

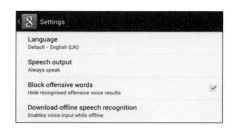

5 Tap on the **Text-to-speech output** button in Step 3 on page 57 to make selections and tap here for specific settings

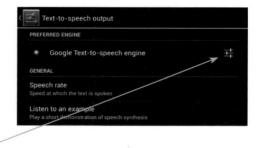

Backup & reset

1 Tap on the **Backup & reset** button under **Personal**

2 Tap on the **Back up my data** checkbox to ensure that your data and settings are backed up by Google

3 Tap on the **Factory data reset** button to remove all data and return the tablet to its orginal state

Accounts

The Account settings can be used to add both email and Google accounts to your tablet.

Add account

① Tap on the **Add account** button under **Accounts**

② Tap on the **Corporate** button to add an account that can be used with an Exchange email client

Don't forget

If you are setting up a corporate account through Exchange you will probably need to obtain the required settings from your IT administrator.

③ Enter the Exchange account details. Tap on the **Next** button on the page or on the keyboard to complete the setup process for Exchange

Don't forget

For some versions of Android there may not be a Corporate option. Instead, there may be options for Microsoft Exchange and Server.

④ Tap on the **Email** button in Step 2 and enter details for a personal email account. Tap on the **Next** button, or the **Manual setup** button to enter server details for the account

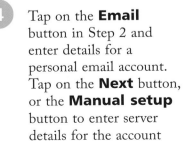

...cont'd

5 Tap on the type of email account that you want to create. This may have to be done if you do not have a webmail account

Don't forget

For more details about using Email on your tablet, see Chapter Nine.

6 Enter the settings for your email account and tap on the **Next** button to complete the setup process

7 Tap on the **Google** button in Step 2 on the previous page. Tap on the **New** button to set up a new Google Account. See Chapter One for details about how to do this

System

The System settings can be used to set the system date and time of the tablet, various accessibility options and information regarding updates to the system software.

Date & Time

1. Tap on the **Date & time** button under **System**

2. Tap on this checkbox to set the date and time automatically

3. Tap on the **Select time zone** button to set a specific time zone

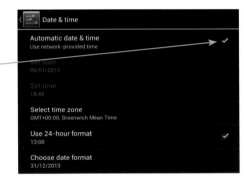

Accessibility

1. Tap on the **Accessibility** button under **System**

2. Tap on the accessibility options to activate them or access their individual settings and options

Don't forget

The current time appears on the Lock screen of your tablet.

Don't forget

Some versions of Android do not have the automatic date and time option.

...cont'd

About tablet

This can be used to update the system software on your tablet, i.e. the version of the Android operating system. To do this:

Hot tip

The current version of Android on your tablet is displayed under the **Android version** heading in the **About tablet** section.

Don't forget

On some tablets, the **About tablet** option is called **About device** and the **System updates** option is called **Software update**.

1 Tap on the **About tablet** button under **System**

2 Tap on the **System updates** button

About tablet

System updates

Status
Status of the battery, network and other information

Legal information

FCC ID: MSQME370T

IC: 3568A-ME370T
Model: ME370T

Model number
Nexus 7

Android version
4.2.1

3 If you have the latest available version of Android for your tablet this will be noted here. If there is an update available you will be able to download it

Your system is up to date.

Last checked for update at 14:24.

4 Updates to Android are checked for automatically but you can also check manually at any time by tapping on the **Check now** button

Check now

4 At Your Fingertips

Although there is no physical keyboard with an Android tablet, the touchscreen one offers good functionality for entering and working with text and data using various keyboard functions.

The Android Keyboard

The keyboard on an Android tablet is a virtual one, i.e. it appears on the touchscreen whenever text or numbered input is required for an app. This can be for a variety of reasons:

- Entering text with a word processing app, email or an organizing app

- Entering a web address

- Entering information into a form

- Entering a password

Viewing the keyboard

When you attempt one of the items above, the keyboard appears before you can enter any text or numbers:

Around the keyboard

To access the various keyboard controls:

1 Tap once on this button to add a single capital letter (the keyboard letters are displayed as capitals)

2 Double-tap on this button to create **Caps Lock**. This is indicated by a blue line underneath the arrow

3 Tap once on this button to back-delete an item

4 Tap once on this button to access the **Numbers** keyboard option

5 From the Numbers keyboard, tap once on this button to access the **Symbols** keyboard

65

6 Tap once on this button on either of the two keyboards above to return to the standard **QWERTY** option

7 Tap once on this button to hide the keyboard (this can be done from the Navigation Bar at the bottom of the screen). If the keyboard is hidden, tap once on one of the input options, e.g. entering text, to show it again

Keyboard Settings

There are a number of options for setting up the functionality of your Android tablet's keyboard. These can be accessed from the Personal Settings:

1 Open the **Settings** app and tap on the **Language & input** button under **Personal**

2 Tap here to enable the **Spell checker**

3 Tap here to access the settings for the default Android keyboard

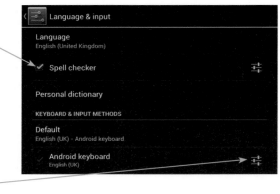

4 Tap here to activate **Auto-capitalisation** so that the first word of each new sentence automatically starts with a capital

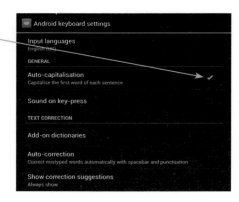

5 Tap on the **Add-on dictionaries** button to add more dictionaries to the keyboard

6 Tap on the **Auto-correction** button in Step 4 and tap on a radio button for the level of auto-correction

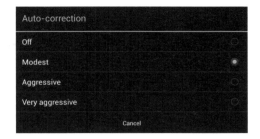

7 Tap on the **Show correction suggestions** button in Step 4 and tap on a radio button to determine how auto-correction suggestions are shown

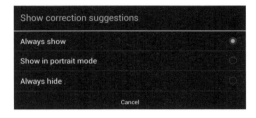

8 Tap here to enable **Next word suggestions** for auto-correction, which helps to make the suggestion in the context of what is being written

9 Tap on the **Advanced settings** button to make selections for using suggestions from your

People app (Address Book) when you start typing a name, showing when other languages can be used on the keyboard and switching to another input method

Keyboard Shortcuts

Because of the size of the keyboard on an Android tablet, some keys have duplicate functionality, in order to fit in all of the options. This includes keys with dual functions, accented letters and keyboard settings options.

Much of this functionality is accessed by pressing and holding on keys, rather than just tapping on them once.

Keyboard settings

To access the keyboard settings directly from the keyboard:

Beware

If **Google voice typing** is turned off (**Settings > Language & input** and check off **Google voice typing**) the button in Step 1 shows the **Settings** button, rather than the **Microphone** one.

 Press and hold on this button

 Tap on the **Settings** button

Tap on **Input languages** or **Android keyboard settings** to access the full range of settings for the keyboard

Input options
Input languages
Android keyboard settings

Dual functions

If a key has more than one character on it, both items can be from the same button.

Tap on a button to insert the main character

Press and hold on the button to access the second character (in the top right-hand corner on the key). Release the button to insert the character

Accented letters

Specific letters on the keyboard can also be used to include accented letters for words in different languages.

1 Press and hold on a letter that has corresponding accented versions in different languages. Tap on a letter to insert it

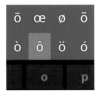

2 The **Return** key also has a **Previous** button that can be accessed by pressing and holding on it. This can be used when filling in online forms

Spacebar shortcuts

The spacebar can also be used for a useful shortcut:

1 At the end of a sentence, double-tap on the spacebar to add a full stop/period and a space, ready for the start of the next sentence

He left the building.

Gesture Typing

Android 4.2, and higher, supports gesture typing, which means you can create text by swiping over letters rather than tapping on individual keys on the keyboard. This can be enabled within the keyboard settings section. To do this:

1 Access the **Settings** app and under **Language & input**, tap on the **Android keyboard** settings button

2 In the **Gesture Typing** section, tap on the checkboxes to enable gesture typing and its functionality

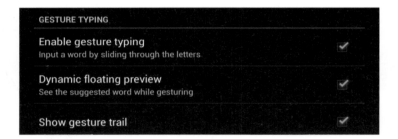

3 Swipe between the required letters on the keyboard to create the required words. If the **Show gesture trail** box is checked on in Step 2, a blue line will show the path of your finger on the keyboard

4 The word appears underlined in the app. Tap on the spacebar to include it

Beware

Gesture typing is best suited to short messages, such as text messages or short emails, rather than trying to write lengthy documents with it.

Adding Emojis

They go by a variety of name (smileys, emoticons, emojis) and tend to split opinion along the lines of, 'silly nonsense' or 'great fun'. Whatever your views on them, it is possible to add emojis on the keyboard of an Android tablet. To do this you have to first add the emojis as an add-on dictionary:

1 Access the keyboard settings as shown on the previous page and tap on the **Add-on dictionaries** button

2 Tap on the **Emoji for English words** button

3 Tap on the **Install** button

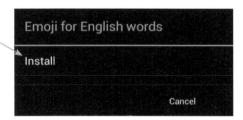

4 Once the Emoji dictionary has been installed this is noted underneath its name

5 On the keyboard, the **Emoji** button is available in the bottom right-hand corner

6 Press and hold on the **Emoji** button and tap on a symbol to add that to your text

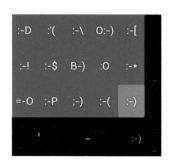

Beware

Use emojis sparingly as they can become rather annoying for the recipient of them.

Adding Text

Once you have applied the keyboard settings that you require you can start entering text. To do this:

72

1 Tap once on the screen to activate the keyboard. Start typing with the keyboard. The text will appear at the point where you tapped on the screen

2 If **Auto-correction** is enabled, suggestions appear above the keyboard. Tap once on the spacebar to accept the suggestion with the dots underneath it, or tap on another word to insert that instead

3 If the **Spell checker** is enabled, any misspelled words appear underlined in red

Once upon a tyme|

4 Tap on a misspelled work to view suggested alternatives. Tap on one to replace the misspelled word, or add it to the dictionary, or delete it

Working with Text

Once text has been entered it can be selected, copied, cut and pasted, either within an app or between apps.

Selecting text

To select text and perform tasks on it:

1 Tap anywhere to set the insertion point for adding, or editing, text

2 Drag the marker to move the insertion point

3 Double-tap on a word to activate the selection handles

4 Drag the handles to change the text that is selected

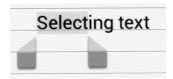

5 Tap on these buttons at the top of the window to **Cut** or **Copy** the selected text

6 Locate the point at which you want to include the text. Press and hold and tap on **Paste** to add the text

Don't forget

There is usually a **Paste** option on the main toolbar of apps in which you want to paste text that has been copied.

Creating a Dictionary

In addition to the standard dictionaries on your Android tablet it is also possible to create your own custom dictionary. This could include real names, or words which you use regularly but do not appear in the virtual dictionary. To create your own dictionary:

 Access **Settings > Language & input** and tap on the **Personal dictionary** button

Beware

Do not use a shortcut that is an actual word, otherwise is could cause confusion.

At this point the dictionary will be empty. Tap on the **Add** button to create an entry

Enter the word, or phrase, and a shortcut with which to enter it. Tap on the **Done** button on the keyboard

The entry is added to your custom dictionary

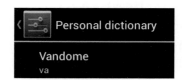

When entering text, enter the shortcut. The word will appear on the options bar above the keyboard. Tap on the required word to insert it

5 Working with Apps

Apps on an Android tablet are the programs that provide the functionality for all of its operation. This chapter details the built-in ones and shows how to obtain many more.

About Android Apps

An app is just a more modern name for a computer program. The terminology first became widely used on smartphones, but it has now spread across all forms of computing and is firmly embedded in the vocabulary of tablets.

On Android tablets there are two types of apps:

- **Built-in apps.** These are the apps that come pre-installed on your Android tablet

- **Play Store apps.** These are apps that can be downloaded from the online Play Store

Using apps

To use apps on an Android tablet:

1 The full range of Android apps can be viewed by tapping on the **All Apps** button

2 Tap once on an app to open it

3 Tap on the **Menu** button to access individual settings for a specific app

4 Individual apps also have toolbars with buttons for accessing the functionality for the app

Built-in Apps

The built-in apps that are available from the All Apps screen can vary slightly depending on the brand of Android tablet being used. However, some of the generic apps should include:

- **Calculator.** A standard calculator that also has some scientific functions, although not the range of a full scientific calculator

- **Calendar.** An app for storing appointments, important dates and other calendar information. It can be synced with your Google Account

- **Chrome.** Different Android tablets have different types of browsers for accessing the Web. The Chrome browser is the default on some tablets

- **Clock.** This can be used to view the time in different countries and also as an alarm clock and a stopwatch

- **Currents.** This is an app that displays the latest news from a selection of online media outlets. Additional services can be added to the defaults

- **Downloads.** When you download content onto your tablet, such as from a website or an email, this can be viewed and managed in this app

- **Earth.** This is the Google Earth app that provides photographic images of the globe. You can zoom in on items and also tilt the view

- **Email.** This can be used to manage all of the email accounts on your tablet. This includes your Gmail account and any other accounts you have added

...cont'd

● **Gallery.** This is the default Android app for viewing, managing and sharing your photos

Gallery

● **Gmail.** When you set up a Google Account you will also create a Gmail account for sending and receiving email. This app can be used for this

Gmail

● **Google.** This app can be used for accessing the Google search function, still one of the best search facilities available. It can also be used for access the Google Now function, if set up

Google

Don't forget

Several of the Google apps are interlinked, i.e. the Local app uses Google Maps, and Google Maps can also be used to find local information.

● **Google+.** This can be used with your Google Account to share content such as photos and updates with specific people

Google+

● **Local.** This is another Google app and provides location-specific information about restaurants, cafes and local attractions

Local

● **Maps.** Google Maps app is one of the best mapping apps available for finding locations and obtaining directions. It can also be used in conjunction with the Local app

Maps

● **Messenger.** This is an app for tablets and mobile devices that can be used to have text or video chats with groups of friends and also share content such as photos and videos

Messenger

● **Navigation.** Another app that uses Google Maps and offers directions and also traffic information using your current location via GPS

Navigation

● **People.** This is the Android address book where you can enter details about your friends, family members and business contacts. It can also be synced with your Google Account

People

- **Play Books.** This is the app for reading ebooks on an Android tablet. It can be used to manage the books that you have downloaded and also access the books section of the Play Store

- **Play Magazines.** Similar to the Play Books app, this is used for reading magazine and downloading new ones from the Play Store

- **Play Movies.** Another app linked to the Play Store, this is used to view movies that you have bought or rented from the Play Store. It can also be used to view your own personal videos

Don't forget

Some Android tablets have different apps for functions such as playing music and movies and reading books and magazines. If this is the case, the Play ones here can still be downloaded from the Play Store.

- **Play Music.** This is the default music player that can be used to play your own music and also music content from the Play Store

- **Play Store.** This is where all of the online content for Android tablets can be accessed, bought and downloaded. This includes apps, books, music, movies and magazines

- **Settings.** This is where all of the Android settings can be accessed, as shown in Chapter Three

- **Talk.** This is a Google chat app that can be used for text and voice chatting over the Internet. It can also be used for transferring files and photos

- **Voice Search.** This can be used for search queries using speech rather than typing. It can also be used to open apps or perform tasks such as finding direction via the Maps app

- **YouTube.** This provides direct access to the popular video sharing website

Finding Apps

Although the built-in apps provide a lot of useful functionality and are a good starting point, the Play Store is where you can really start to take advantage of the wide range of apps that is available. These can be used for entertainment, communication, productivity and much more.

To access the Play Store and find apps:

1   Tap on the **Play Store** app

2   Suggested items are shown on the Play Store homepage

3   Swipe up and down to see the full range of recommendations. Tap on an item to view further details about it

Don't forget

New apps are added to the Play Store on a regular basis (and existing ones are updated) so the home page will change appearance regularly too.

4 Use these buttons to find content according to specific categories

5 Tap on this button to search for specific items

6 Enter the name of the item for which you want to search. Tap on one of the suggested results or tap on this button on the keyboard

7 Tap on the **Menu** button to access the additional options including viewing information about your Google Account and also Settings for the Play Store

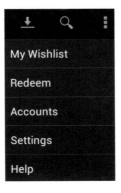

> **Hot tip**
>
> If you have a Play Store Gift Card, it can be redeemed from the **Redeem** button in Step 7. Enter the Gift Card code and the relevant amount will be credited to your Google Play balance for use in the Play Store.

Downloading Apps

Once you have found an app in the Play Store that you want to use, you can download it to your tablet:

1 Access the app you want to use. Tap here to view a video preview of the app (if available). There will also be details about the app and reviews from other users

2 Tap on the **Install** button

Beware

If updates are set to be uploaded automatically you will have less control over what is added to your tablet. Also, an active Internet connection is required for automatic updates.

3 Tap on the **Accept & download** button

4 You will be notified that the app is downloading. Tap on this checkbox if you want future updates to be uploaded automatically

5 Newly downloaded apps are added alphabetically where there is space on the next available Home screen

6 For a paid-for app you have to check on the **Payments** checkbox and tap on the **Accept & buy** button

Deleting Apps

The built-in apps on an Android tablet cannot be deleted easily, but the ones that have been downloaded from the Play Store can be removed. You may want to do this if you do not use a certain app any more and you feel the number of apps on your tablet is becoming unmanageable. To delete a downloaded app:

Tap on the **Settings** app

Under **Device**, tap on the **Apps** button

Tap on the **All** tab and tap on an app

Tap on the **Uninstall** button to remove the app

Tap on the **OK** button to confirm the removal

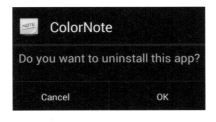

Hot tip

Built-in apps have a **Disable** button instead of an **Uninstall** one in Step 4. Tap on this to stop the app's operation. This does not remove it and it can be reinstated with the corresponding **Enable** button.

Updating Apps

The world of apps is a dynamic and fast-moving one and new apps are being created and added to the Play Store on a daily basis. Existing apps are also being updated, to improve their performance and functionality. Once you have installed an app from the Play Store it is possible to obtain updates, at no extra cost (if the app was paid for). To do this:

1 Access the Play Store and tap on this button

2 Tap on the **Installed** tab to view the apps on your tablet. Tap here to update all of the appropriate apps or tap on the **Update** button next to a specific app to update it

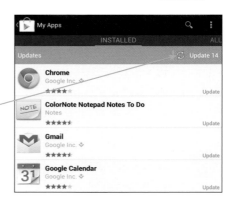

3 Tap on the **Update** button to download the latest version of the app, including any enhancements

4 Tap on the **Accept & download** button and check here for the permissions that the app is asking to use

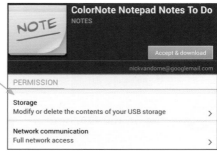

5 The update is displayed on the **Notifications Bar**

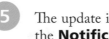

App Information

For both built-in apps and those downloaded from the Play Store it is possible to view details about them and also see the permissions that they are using to access certain functions. To view information about your apps:

1   Open the **Settings** app and tap on the **Apps** button under **Device**

2   Tap on the **All** tab and tap on an app

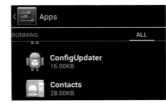

3   Tap on the **Force stop** button to close a running app

4   Details about the size of the app, and the amount of data it has stored, is shown here

5   Swipe down to **Permissions** to view what the app is accessing on your tablet

Don't forget

Check on the **Show notifications** checkbox to have alerts and updates for the app shown in the Notifications Bar.

Beware

The **Clear data** button removes all of an app's settings and data that it has stored. The **Clear cache** button removes data that has been stored temporarily in the app's memory.

Becoming a Developer

If you have a good programming knowledge of Android you may want to work as a developer in terms of producing your own apps. In early versions of Android it was possible to access the developer options from the About button under System. However, in Android 4.2, and later, this has been hidden, although it is still available. To access it:

1 Open the **Settings** app and tap on the **About tablet** button under **System**

2 Tap several times on the **Android version** button

3 After a few taps you will be alerted that you are accessing the developer options. The Android screen appears which indicates that the developer options have been unlocked

4 Under **System** the **Developer options** button is now available

5 Tap on the **Developer options** button to view them. These options should only be used if you are familiar with the Android API interface

6 Using Google Now

Google Now is a service that gives you up-to-date information in real time. This chapter shows how to set it up and use it to get updates on your favorite topics.

About Google Now

We live in an age where we want the availability of as much up-to-date information as possible. On an Android tablet, one option for this is Google Now. This is an online service that provides items such as the latest traffic information for your area, flight information or the results from your favorite sports team. The information displayed, and when it is displayed, is tailored to your needs according to the apps that you use and the type of content that you access.

When Google Now is activated this also turns on your location history so that Google can make use of the location data that is collected by your tablet.

Google Now cards

The functionality of Google Now is provided by cards. These display up-to-date information for a variety of topics. You can apply your own specific settings for each card and these will then display new information as it is occurs. Sample cards can be viewed and the individual settings for each card can also be accessed from here. Once a card has been set up you do not have to do anything else; just wait for the cards to appear with updates.

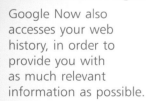

88

1:07 to Work
Normal traffic on M90

Perth
St. Andrews
Glenrothes
Kirkcaldy
Dunfermline
Glasgow
Edinb

Navigate

Bulls   100
Celtics   99
Final, 19 January

Recap & Highlights

Box score

Accessing Google Now

By default, Google Now is not on. To activate it so that it works for you in the background:

1 Swipe up from the bottom of the screen

2 If the keyboard is showing, tap on this button to hide it

3 Tap on this button in the bottom right-hand corner

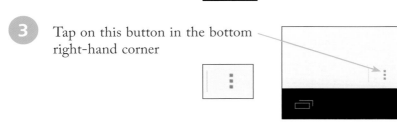

Don't forget

Google Now can also be accessed from the **Google** app in the **All Apps** section. Open this and continue from Step 3.

Google

4 Tap on the **Settings** button

Settings
Refresh
Send feedback
Help

5 Tap on the **Google Now** button

6 Tap on the **Yes, I'm in** button

Get Google Now!
Google Now is always working for you. It needs to:
· Use and store your location periodically for traffic alerts, directions and more.
· Use your synced calendars, Gmail and Google data for reminders and other suggestions.

Learn more

No, maybe later.   Yes, I'm in.

Around Google Now

When you first activate Google Now you will see the Home screen. This is where your cards will show up and from where you can access all of the Google Now Settings, for turning cards on and off and applying your own settings to them.

1 Active cards are shown on the Home screen

2 Tap here to view the range of sample cards

3 Tap here on a card and tap on the **Settings** button to view the card's individual settings

4 Swipe to the bottom of the page and tap on the **Show sample cards** button to view these, or tap here to view the full range of settings (see next page)

5 Tap on a card to view more details about the subject in a related app, i.e. on the Web (if applicable)

Google Now Settings

The full range of Google Now Settings can be used to turn cards on or off for different topics and also apply your own criteria for how they operate. To do this:

1 Tap on this button at the bottom or the screen

2 Tap on the **Settings** button

Settings

3 Tap on the **Google Now** button

Hot tip

To disable Google Now, drag this button to **Off**.

4 The full range of available cards is displayed. Tap on a card title to access its specific settings

Google Now

Weather
Always, Where you are, For upcoming trips

Traffic
Before commute from home, Before commute to home, Before other commutes, After searching Google, Airport/hotel if travelling, Other places if travelling

Next appointment
For any meeting

Travel
With a different currency, With a different language, In a different time zone, With popular attractions

Flights
After flight search

Public transport
Around train stations, Around bus stops, When travelling

Places
Restaurants/bars, Museums, Shops

Sports
Before a game, During a game, After a game

Public Alerts
For emergency messages

Movies
On your movie days, When near a cinema, For new movie you may like

Stocks
On 1% price change or more

News update
When a news story you are interested in has been updated

Hot tip

Swipe to the bottom of the **Settings** window to select **Notification** options for when there is an update to a card. This will appear on the Notifications Bar and you can also specify ringtone and vibration options.

5 Drag this button to **On** to activate a card

6 Tap on an item to enter your own criteria for its settings

Weather

Card appears
Always

Weather location
Where you are, For upcoming trips

Weather units
Celsius

NOTIFICATIONS

Weather when card appears
Off

Using Your Web History

A lot of the information for Google Now cards is taken from your related Web History. This is in relation to searches that you have performed using Google, or items that you have accessed using Google products, such as Google Maps. The Google Now cards use this information to tailor the results that are shown on the cards to make them as specific to you as possible.

Your Google Web History is different from the standard web history stored by a browser and is stored within your Google Account. This has to be turned on to enable the full functionality of Google Now. To do this:

Hot tip

Your Google Web History is collected from a variety of sources, not just from web pages. If you delete your web browser's history, this will not affect your Google Web History.

1 Access the web page at **google.com/history/** and sign in to your Google Account

2 On the Web History page, tap on the **Turn Web History on** button to activate this

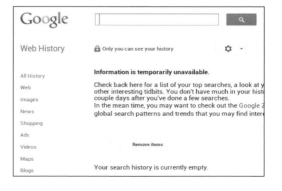

3 Initially the Web History is empty. When you start performing searches and visiting web pages these pages will appear here

4 Once you have done some searches over a few days your Web History starts to be built up. This includes the times and dates on which you performed most searches, the number of searches per day and also a list of specific items that have been searched for

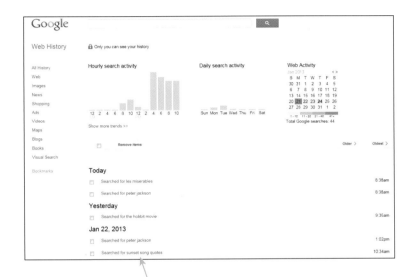

5 The searches are displayed here and are used by Google Now to tailor the information that you are shown on the cards. For instance, the Movies card may use this information to display details of new movies, based on the information for which you have already searched, such as movie stars or directors

6 If you do not want an item to be shown in your Web History, and used by Google Now, check on the checkbox next to it and click or tap on the **Remove items** button

Google Now Cards

There are two types of Google Now cards: the standard ones and those that are linked to information in your Gmail account, such as restaurant or flight bookings that have been made with apps linked to Gmail.

The standard Google Now cards are generally linked to your location and the information obtained from your web searches.

Weather

This card can be used to display the four-day weather forecast for your current location, your home location, your work location or for any booked trips.

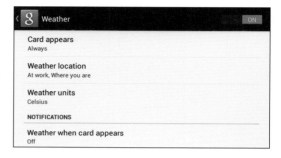

Don't forget

The settings for the Weather card can also be used to specify whether the temperature is in Celsius or Fahrenheit.

The card can be set to appear in the morning, the evening, or always.

If more than one location is selected, they are shown stacked on top of each other. Tap on the one at the back to view it.

94

Traffic

This card can be used to show the traffic conditions for your daily commute. It can be set to appear before your commute from home, before you commute to home, before other commutes, after searching Google and when you are traveling in other locations. You can also select a transportation mode of driving or public transport and specify notifications for specific traffic conditions.

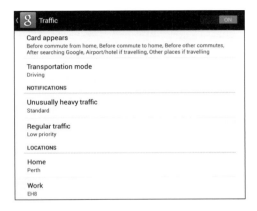

Don't forget

Tap on the items under the **Locations** setting to set your home location and your destination. The destination can be given a specific name.

95

On the card, tap on the **Navigate** button to get directions to your destination. This is done through the Navigation app which provides directions via GPS.

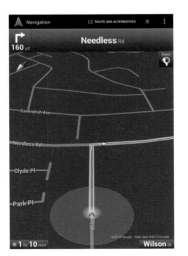

Hot tip

When you enter a new destination search into the Maps app, this shows up on the Google Now Traffic card.

...cont'd

Next Appointment

This card can be used to display the next event in your calendar. It can be set to appear for any meeting, only if it is far away or in the evening for a meeting that is the next day.

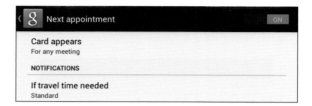

When appointments, or events, are added to the Calendar app the next one is displayed in Google Now.

Travel

This card is dependant on your geographical location. If it is turned on, your tablet's location access services will determine your location in the world and provide information accordingly in Google Now. This includes local currency, translation options, the time back home and information about local attractions.

Flights

This card appears with relevant details after you have conducted a search on the Web for flights. The options are for just checking on or off the **After flight search** checkbox.

Public transport

When you are near to bus or train stations this card will appear with details of available public transport. The options are for showing the card when around train stations, bus stops or when traveling anywhere.

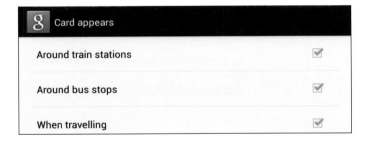

Places

This card can be used to show places of interest when you are in a town or city. The options are for showing the card with relevant items when near to restaurant/bars, museums and shops.

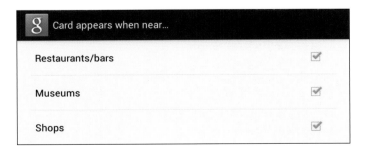

...cont'd

Sports

This card can be used to view the upcoming games, and results, of your favorite teams. You can select the teams that you want to appear on the card. The card can appear before, during and after games, or all three.

To select a team to add to the Sports card:

98

1 Tap on the **Add team** button below the **Teams** setting

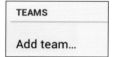

2 Tap in the **Add team** box

3 As you start typing, suggested teams appear. Tap on a team name to add it

4 Details of the team's games are displayed on the Google Now Home screen. If there is no game taking place there may be a preview of the next one. If it is the off-season for the selected sport there may not be anything displayed

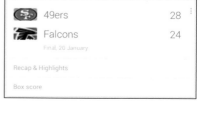

Public Alerts

This card can be used to view details of public service messages relating to natural disasters such as storms and earthquakes. However, it is only available in areas where this type of service is provided, usually the US.

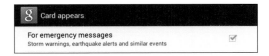

Movies

This card can be used to display movie information for your area, based on the days when you search for movies, when you are physically near a cinema and for new movies based on your web searches for movies, artists and directors.

Don't forget

A lot of the Google Now cards operate by accessing your Web History, which can be viewed using your Google Account details on the Google website. For more information, see page 92.

The information changes according to which movies are currently playing near your location.

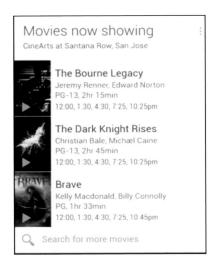

Stocks

This card can be used to display up-to-date stock market information. You can select stocks in the same way as for selecting sports teams and the card can be displayed for always, mornings and

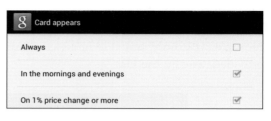

evenings, or when there is a 1% change in the price of one of your selected stocks.

Information displayed on the Stocks card is usually delayed by approximately 15 minutes from the real-time price.

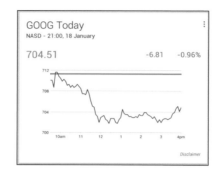

News update

This card can be used to display updated, or related, news items in relation to those that you have looked at on the Web.

Photo spot nearby

This card can be used to display good photo opportunities when you are out and about. An active Internet connection is required for this so that location-services can identify where you are and locate the nearest photo spots that are identifed by Google.

Events

This card can be used to display local events according to your current location.

Birthdays

On your birthday, this card will display a personalized message. It can also be used to display the birthday of your contacts if you are signed in to Google+ account.

Research topics

This card displays information related to any research that you have done on the Web for specific topics, such as a location that you want to visit, or a subject for a project.

Activity summary

This card provides a monthly summary of approximately how far you have walked or cycled during that period. To use this you have to take your tablet with you whenever you are walking or cycling and have an Internet connection so that the location data can be collated by Google.

Gmail Cards

Gmail cards are ones that are linked to your Gmail account (this is the email account that is created when you sign up for a Google Account). This is for items that are booked or bought online and the confirmation email is sent to your Gmail account. Not all companies participate in this so it can be a bit hit and miss in terms of what shows up in the Gmail cards. The range of available Gmail cards are:

Packages
This card can be used to display information about deliveries for items that you have bought online and which use Gmail for the confirmation email. On the card there is a link to track the progress of the package.

Flights
This card can be used to display information about flights that you have bought online and which use Gmail for the confirmation email. On the card is a link containing navigation details to the airport where the flight leaves from and the approximate amount of time that it will take to get there from your current location.

Hotels
This card can be used for hotel bookings that have been made online and which use Gmail for the confirmation email. On the card is a link for directions to the hotel.

Restaurants
This card can be used for restaurant bookings that have been made online and which use Gmail for the confirmation email. On the card is a link for directions to the restaurant.

Event bookings
This card can be used for events that have been booked online and which use Gmail for the confirmation email. On the card is a link for directions to the event.

Barcodes
This card can be used to display information about items containing a barcode that have been bought online and which use Gmail for the confirmation email. Most commonly this is for airline boarding passes.

Don't forget

For each Gmail card there is a button to view the related email for the card, i.e. the one from which the card is generated.

7 Getting Musical

Listening to music is an integral function of a mobile device and this is well catered for on Android tablets. This chapter shows how to buy, download and transfer music and options for playing it too.

Using Google Play

Google Play is the online store for buying, downloading, using and managing a range of entertainment content. It is accessed at the website:

- **play.google.com**

You need to have a Google Account in order to log in to the Google Play website.

Don't forget

Content that is downloaded to your Android tablet via the Play Store will also be available on the Google Play website, as long as you are logged in with your Google Account.

Once you have logged in to Google Play you can download a variety of content:

- Music

- Books

- Magazines

- Movies

- Apps

Content from Google Play is stored in the cloud so it can then be used on your computer and also your Android tablet. If you delete it from your tablet, either accidentally or on purpose, you can still reinstall it from Google Play. You can also use content downloaded by any other Android devices, such as a smartphone.

Music on Android

One option for playing music on an Android tablet is the Play Music app. It can be used to play music that has been obtained in a number of different ways:

- Downloaded directly to your tablet from the **Play Store**

- Bought on **Google Play** and then used on your tablet

- Uploaded to **Google Play** and then downloaded to your tablet. This can be done with the **Music Manager** that can be installed directly from Google Play

- Transferred from your computer directly to your tablet. This is done by connecting your tablet to your computer, using the USB cable, and copying your music to the **Music** folder on your tablet

- Transferred from another mobile device using Bluetooth

Don't forget

Some tablets have their own default music app, which will be linked to their own app store. If this is the case, the Play Music app can still be downloaded from the Play Store and content can be bought from there.

Hot tip

If you buy music from either the Play Store or the Google Play website, it can be played on your tablet with streaming (using a Wi-Fi Internet connection) or it can be saved (pinned) onto your tablet so that you can listen to it offline too.

Don't forget

Whenever you buy music on Google Play, or upload it from your computer, it will appear in the My Music section.

Downloading Music

Downloading from the Play Store

To use your tablet directly to find and download music from the Play Store:

1 Tap on the **Play Music** app to open the music player

2 The music player will be empty if you have not downloaded or transferred any music

3 Tap on the **Play Store** button

4 Use these button to view the relevant sections within the Music section of the Play Store, or

5 Tap on these sections to view the recommended content, or

6 Tap on this button and enter an artist, album or song name into the Search box

7 Locate the item you want to download. For an album, tap on this button to buy the full album or tap on the price button next to an individual song

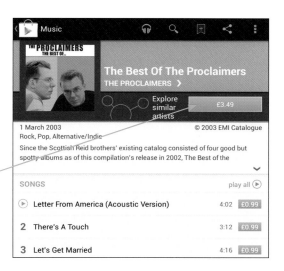

8 Tap on the **Accept & buy** button and check on this box to agree to the purchase

9 Once an item has been downloaded, tap on the **Listen** button to listen to it on your tablet

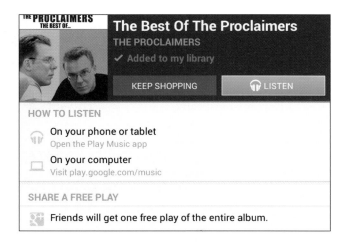

Don't forget

When you buy music from the Play Store it also comes with the related artwork such as album or single covers.

Hot tip

When music is bought through the Play Store app it is also available through the Google Play website when you are logged in.

...cont'd

Downloading from the Google Play website

Music can also be purchased and downloaded to your tablet directly from the Google Play website:

1 Log in to Google Play with your Google Account details

2 Tap on the **Music** button

3 Find the item(s) you want to buy and download and tap on the **Purchase** button

4 Review the purchase details and tap on the **Continue** button

5 Once the item has been purchased it will be available on the Google Play website and also on your tablet. Tap on the **Go To My Music** button to access it on Google Play

Transferring Music

Music does not have to be bought from the Play Store or the Google Play website in order to listen to it on your tablet. You can use music that you already have and transfer it onto your tablet. This can be done via the Google Play website or directly from your computer.

Transferring via Google Play

Your own music can be uploaded to the Google Play website and it will then become available on your tablet.

1. Log in to Google Play with your Google Account details

2. Click on the **My Music** button

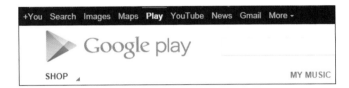

3. Click on the **Upload Music** button

4. The Music Manager app is used for adding your music to Google Play. Click on this button to download it

The Music Manager app is quite small in terms of size and should not take too long to download to your computer.

5. On your computer, click on this button to open the **Music Manager**

...cont'd

6 The Music Manager
automatically scans
your music folders
and uploads them.
Click on the **Add
Folder** button to
add music from
another location and click on the **OK** button

7 The progress of the
upload is shown
in the left-hand
panel. Check on this
checkbox to have
new music uploaded
automatically when
it is added to the
selected music folders

Beware

The Music Manager does
not support DRM (Digital
Rights Management)
AAC files, which are
often used in iTunes.
These files will have
to be converted into
another format, for
example, MP3, before
they are uploaded by the
Music Manager.

8 If any songs cannot be
uploaded this is indicated
in the upload panel.
Click on the link that
states how many songs
cannot be uploaded to
view these details

9 Once the music
is uploaded it is
available under the
My Music section
of your Google Play
account. It is also
made available on
your tablet

110

Transferring from a computer

If you want to transfer music from your computer to your Android tablet, this can be done by connecting the two devices with the USB micro connector cable supplied with the tablet. To do this:

1 Connect the tablet and the computer with the cable

2 Browse to the **Music** folder on your computer. This should be where your music is stored

3 Select an item and copy it

4 Locate your tablet in your computer's file manager

5 Browse to the **Music** folder of the tablet

6 Paste the copied item. A warning window appears alerting you to the fact that the file format may not be compatible with your tablet. Click on the **Yes** button

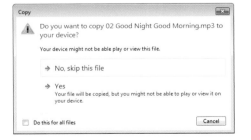

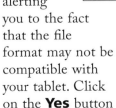

Beware

To transfer music by Bluetooth this has to be enabled on both devices.

Playing Music

Once you have obtained music on your tablet, by whatever means, you can then start playing it and listening to it. To do this:

1 Tap on the **Play Music** app

2 All of the available music is displayed. This includes music from the Google Play website that is only available for streaming at this point, i.e. it needs a Wi-Fi connection to play it

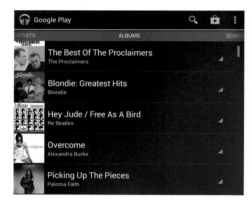

Don't forget

The latest music to be played is displayed in the **My Library** widget. Tap on an item here to open it directly in the **Play Music** app.

112

3 Tap on these buttons to view your music content according to **Playlists**, **Recent**, **Artists**, **Albums** and **Songs**

4 Tap on an item to view the available songs (for an album)

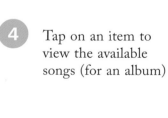

5 Tap on a song to play it. This is displayed in the music player, with any related artwork for the song or album. The standard playback controls are displayed at the bottom of the window

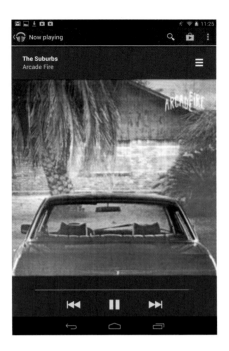

Hot tip

Invest in a reasonable set of headphones to listen to music on your Android tablet. This will usually result in a higher quality than through the built-in speakers.

113

6 When a song is playing, this button appears next to it

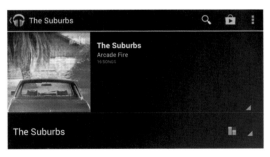

7 The currently-playing song is also displayed at the bottom of the Play Music app window

...cont'd

Music controls

When a song is playing there are several options:

1 Use these buttons to, from left to right, go back to the beginning of
a song, Pause/Play a song, go to the end of a song, i.e. start playing the next one in your music library

2 For an album, tap on this button to view the current queue of songs

Don't forget

Queued songs are those waiting to be played in the music player.

3 Tap in the middle of the screen to access other controls

4 Drag this button to move through a song

5 Tap on this button to shuffle the song in your collection

6 Tap on this button to loop the currently-queued songs

7 Tap on this button again to loop the currently-playing song. This is indicated by a 1 on the button

8 Tap on these buttons to rate the song on the Google Play website

Using the Equaliser

If you want to take a bit more control over the quality of your music, you can use the settings in the Equaliser in the Play Music app. This can be used to make adjustments to the bass and treble sound, using presets or manually. To do this:

1 Start playing a song in the **Play Music** app

2 Tap on the **Menu** button

3 Tap on the Equaliser button

4 Drag the buttons to adjust the bass and treble of your music manually, or

5 Tap on this button and select one of the preset music style options

Hot tip

Other Graphic Equaliser apps can be obtained from the Play Store.

Managing Music

When you are playing music there is still a certain amount of flexibility in terms of managing what is playing, and being scheduled to play. This is know as the music queue. To use this to manage your music:

1 Tap on this button next to an album or a song

2 Tap on the **Add to queue** button to add it to the current queue of songs

3 When a song is playing, tap on this button to view the current queue

4 When viewing the current queue, tap on this button

5 Tap on the **Clear queue** button to remove all of the songs from the current queue

6 When viewing the songs in the queue, tap on the same button as in Step 1 and make one of the related selections, including **Remove from queue**

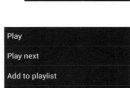

Hot tip

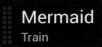

When viewing the current queue, press and hold on the bar to the left of the song title and drag it to reorder its position in the current queue.

Beware

If you select **Clear queue**, this closes the currently-playing item. However, it does not remove it from your tablet and it remains available in the Play Music app.

Pinning Music

Music that is bought from the Play Store or the Google Play website it available for streaming on your tablet using your Wi-Fi Internet connection. This means that the music is sent from the Google servers, where it is stored. This means that it is always backed up and always available.

However, if you are not able to use Wi-Fi you will probably still want to listen to your music, such as when you are traveling. This can be done by pinning the required music to your tablet so that it is physically stored here. To do this:

1 Under one of the main headings, view the available items

2 Tap on this button at the top of the window

3 Check on the **On device only** button to view the items that are stored on your tablet. Those that are just stored in the Google cloud will not be displayed

4 Tap on the button in Step 1 again and tap on the **Choose on-device music** button

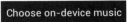

Beware

Items that are not pinned to your device will not be available when you do not have a Wi-Fi connection.

...cont'd

5 The drawing pin buttons denote the status of each item according to whether it is pinned to the device or not. A grayed-out button indicates an item that has been accessed from the Play Store or Google Play website but has not been pinned to the device. A blue button indicates an item that has been transferred directly to the tablet and is automatically pinned

6 Tap on a grayed-out button so that the pin turns white, to start the process for pinning the item to your tablet

118

7 Tap on the tick button at the bottom of the window to save the changes

8 The item will be downloaded for storing on your tablet. This is indicated in the Notifications Bar by the pin icon

9 Access the Play Music app **Settings** (see next page) and tap on the **Download queue** button. The pinned item should be being downloaded, with a pin icon next to it indicating that it will be pinned to the device

10 Once the item has been downloaded it will be available when you select **On device only** as in Step 2 and also when you have no Wi-Fi connection

Play Music Settings

The settings for the Play Music app can be used mainly for managing the music that is downloaded onto your Android tablet.

1 From any window in the Play Music app, tap on this button at the top of the window

2 Tap on the **Settings** button

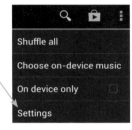

3 The available settings are displayed

4 Tap on this button to see if your Google Account is active

Don't forget

If your Google Account is active this will be denoted by a blue button next to it when you tap on the button in Step 4.

...cont'd

5 Check on this box to hide all items that are not available when you are not connected to the Internet. This is the same as the selecting the **On device only** option

6 Check on this button so that music can be temporarily stored. This is done to improve the downloading process

7 Check on this box to stream music at the highest quality. This is for listening to streamed music, i.e. being downloaded from the Google Play website via Wi-Fi

8 Tap on the **Refresh** button to refresh the music collection from the Google Play website. This can be useful if you think a download process has got stuck

| Refresh |
| Refresh music from Google Play |

Don't forget

The final two settings are for the licences for the Play Music app and its version number.

9 Tap on the **Download queue** button to view the items that are currently being downloaded from the Google Play website or the Play Store

| Download queue |
| Show the songs currently being downloaded |

8 Keeping Entertained

Android tablets are not only very useful; they can also be great fun too. This chapter shows how you can use your tablet as your own personal entertainment center. It details how you can download and watch movies and TV shows, read all of your favorite books and view, manage, edit and share your photos.

Movies and TV Shows

There are different ways in which video content can be viewed on your Android tablet:

- Downloading movies and TV shows from the Play Store or the Google Play website

- Transferring (uploading) your own videos to your tablet

- Watching videos on YouTube

To obtain movies or TV shows from the Play Store:

1. Tap on the **Play Movies** app

2. Tap on the **Movies** tab to view available content. This includes movies and TV shows that you have downloaded and also suggested titles

3. Available movies and TV shows are also shown in the **My Library** widget. Tap on an item here to view it

4. Tap on the **Play Store** button in Step 2 to view the available video content

5 The **Movies** section of the Play Store is similar to those for the other types of content. Tap on these buttons to view movies and TV shows according to these headings

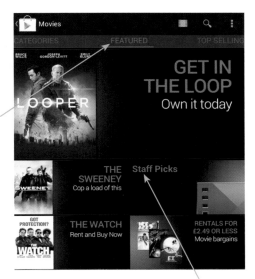

Don't forget

The menu options in the Movies section of the Play Store include adding the currently-viewed title to a Wishlist for buying or downloading later, redeeming a voucher, viewing your account details, viewing general Play Store settings and help options.

6 Tap on the main panels to view highlighted or recommended movies and TV shows

7 Tap on an item to view details about it. Tap here to watch a preview clip

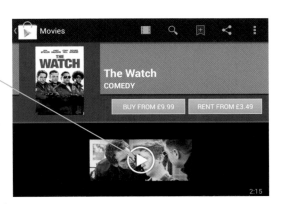

8 Tap on these buttons on the top toolbar to, from left to right, go back to the Play Movies app, search for content in the Play Store, bookmark the item currently being viewed, share an item and access the Movies menu

9 Tap on these buttons to buy or rent a movie or TV show. This will be made available within the Play Movies app and also in the My Library widget

123

...cont'd

10 Tap on the **Accept & buy** button to accept the Play Store terms and conditions

124

11 Enter your Google Account password and tap on the **OK** button to complete the purchase

12 Tap on the **Play** button to start playing your purchase. Tap on the **Keep Shopping** button to look for more movies and TV shows

Watching movies and TV shows

When you have bought, or rented, movies or TV shows you can then watch them on your tablet:

1 Open the **Play Movies** app and view your rented and bought content under the **Movies** tab. Tap on the item you want to view

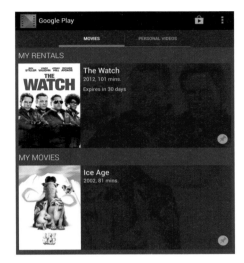

Hot tip

Use an HDMI cable with your tablet so that you can watch movies and TV shows on a High Definition TV.

125

2 Tap on the **Play** button to starting watching a movie or TV show

3 For rented items you have 48 hours to watch them after

Do you want to begin playback? This rental will expire 48 hours after you start watching it.

Cancel   OK

you start watching. Tap on the **OK** button to start

Beware

If you download movies and TV shows to your tablet they can take up a considerable amount of storage space.

4 If the pin icon next to item is grayed-out, the content is streamed from the Play Store. Tap on the pin so that it turns white to download the content to your tablet so that you can watch it offline

Transferring Videos

If you have recorded your own videos on a video camera or smartphone, these can be transferred to your tablet for viewing here. To do this:

1 Connect your tablet to your computer using the supplied USB cable

2 In your file manager, locate the video file that you want to use and copy it

Videos library
Sample Videos
Trees   Water

3 Locate your tablet's **Movies** folder and paste the video here. If a windows appears asking if you want to proceed since not all video file formats will be recognized, tap on the **Yes** button

▶ Computer ▶ Nexus 7 ▶ Internal storage ▶ Movies

Trees
00:00:16
1.52 MB

Copy
Do you want to copy Wildlife.wmv to your device?
Your device might not be able play or view this file.
→ No, skip this file
→ Yes
Your file will be copied, but you might not be able to play or view it on your device.
Do this for all files   Cancel

4 Tap on the **Play Movies** app

Play Movies

5 Tap on the **Personal Videos** tab

6 Compatible videos are displayed. Tap on one to play it in the **Play Movies** app

Google Play
MOVIES   PERSONAL VIDEOS
MOVIES
Trees.mp4
23/01/2013 17:29
00:16

Using YouTube

YouTube is one of the great successes of the Internet age. It is a video sharing site, with millions of video clips covering every subject imaginable (and probably some that have not been imagined either). On an Android tablet the YouTube app can be used to access this vast array of video content. To do this:

1   Tap on the **YouTube** app

2   The trending videos are displayed

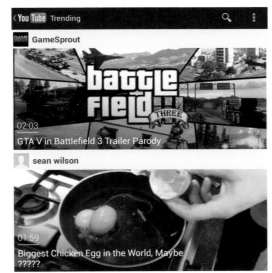

Beware

When you are searching for videos on YouTube it is easy to get sidetracked with other items, due to the vast and diverse amount of content on the site.

3   Tap on the search icon to look for more videos

4   Enter a search word or phrase

5   Matching videos are displayed. Tap on one to start viewing it

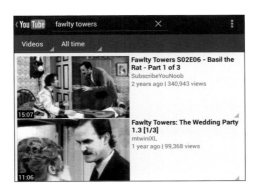

Don't forget

Some tablets have their own default reading app, which will be linked to their own app store. If this is the case, the Play Books app can still be downloaded from the Play Store and content can be bought from there.

Don't forget

The Books section can also be accessed from the main homepage of the **Play Store** by tapping on the **Books** button at the left-hand side of the screen.

Don't forget

Swipe left and right at the top of the window to access the different main headings.

Hot tip

The Kindle app can be downloaded from the Play Store for reading books. If you already have a Kindle account, your books will be available through the Kindle app on your tablet.

Obtaining Books

Due to their size and portability, Android tablets are ideal for reading ebooks. There is a wide range that can be downloaded from the Play Store, or from the Google Play website, in a similar way to obtaining music, movies and apps.

1 Tap on the **Play Books** app (or access the Google Play website)

2 Any books that you already have on your tablet are displayed. Tap on a cover to open a specific title

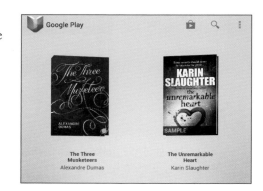

3 Tap on the **Play Store** button

4 Books can be browsed for and downloaded in a similar way as for other Play Store content

5 Tap on these buttons to view books according to these headings

6 Tap on the main panels to view highlighted or recommended books

7 When you find a book you want to read tap on the **Free Sample** button (if there is one) or the **Buy** button (with the price)

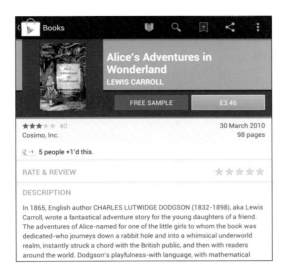

8 The book is downloaded to your tablet and available within your **Play Books** library. Tap in the middle of the page to access the reading controls

ALICE'S ADVENTURES IN WONDERLAND

Around an Ebook

Once you have downloaded ebooks to your tablet you can start reading them. Due to their format there is a certain amount of electronic functionality that is not available in a hard copy version. To find your way around your ebook:

1 Swipe left and right on a page to move backwards or forwards by one page

2 Tap in the middle of a page to access the reading controls toolbars at the top and bottom of the screen

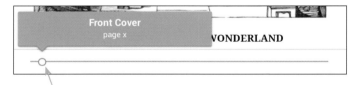

3 Drag this button to move through the book

4 Tap on this button to access the book's table of contents. Tap on a heading to move to that point in the book

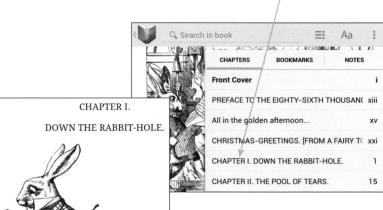

Don't forget

You can also move to the next or previous page by tapping at the right-hand or left-hand edge of a page.

Hot tip

To delete an ebook from the Play Books app, press and hold on its cover and tap on the **Remove from my library** button. The book can be reinstalled from the Play Store if you want to place it in your library again. This is free to do.

5 Tap on the **Menu** button to access the specific settings for the title you are reading

6 Some older books have an option for reading it in its original format. Tap on the **Original pages** button to view this

> ALICE was beginning to get very tired of sitting by her sister on the bank, and of having nothing to do : once or twice she had peeped into the book her sister was reading, but it had no pictures or conversations in it, "and what is

> ALICE was beginning to get very tired of sitting by her sister on the bank, and of having nothing to do: once or twice she had peeped into the book her sister was reading, but it had no pictures or conversations in it, "and what is the use of a book," thought Alice, "without pictures or conversations?"

Don't forget

To go back to the standard Play Books format from the Original pages, tap on the button in Step 5 and tap on the **Flowing text** button.

131

7 Tap on this button to select text options **Aa**

8 Tap on the **Theme** option to select black text on a white background (Day), white text on a black background (Night), or Sepia

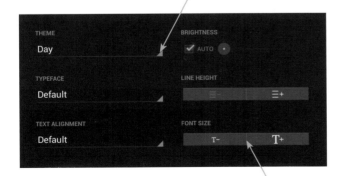

Don't forget

Tap on the **Brightness** button in Step 8 to change the screen brightness for your book.

9 Tap on the **Font Size** buttons to increase or decrease the font size by one step each time. Tap on the **Line Height** buttons to increase or decrease the space between lines on a page

Don't forget

Tap on the **Typeface** option, below the **Theme** one, to select a new typeface for your book. Tap on the **Text Alignment** button to align the text to the left or justify it (flush down the right-hand side).

Adding Notes

If you like taking notes while you are reading books you no longer have to worry about jotting down your thoughts in the margins or on pieces of paper. With Play Books ebooks you can add your own electronic notes and also insert bookmarks at your favorite passages. To do this:

1 Press and hold on a word to activate the two blue text selection markers

same, shedding ga
round her, about f
hall.

After a time she

2 Drag one of the blue markers over the text to select it and access the top toolbar

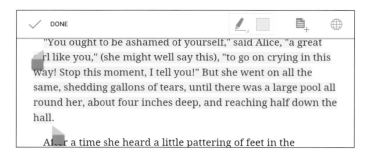

	Yellow
	Green
	Red
	Blue

3 Tap on this button to add a note

4 Enter a note for the selected text and tap on the **Done** button on the top toolbar

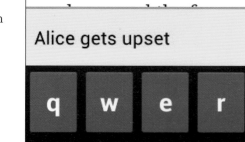

✓ DONE

5 A note is indicated by a small yellow icon next to the highlighted text. Tap on this to edit the note

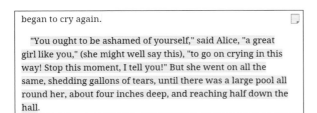

6 To view all of your notes, tap on the **Table of Contents** button and tap on the **Notes** tab

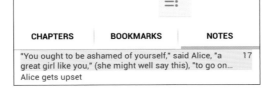

7 Notes are shown on the **Notes** page. Tap on an item to go to that point in the book

8 To delete a note, access it in the book and tap on the yellow note icon

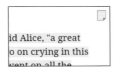

9 On the top toolbar, tap on this button to delete the note

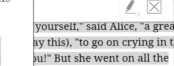

10 Tap on the **Remove** button to confirm the action

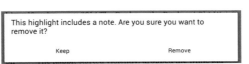

Adding Bookmarks

One of the great things about a hard copy book is that you can physically mark pages, or insert a bookmark, so that you can keep your place. However, with ebooks this functionality has been added so that you can bookmark as many pages as you like, for quick access.

Beware

You can add numerous bookmarks throughout a book. However, if you add too many the **Bookmarks** tab may become cluttered and it will be harder to find items that you want.

1 To bookmark a page in a book, tap in the top right-hand corner. A red bookmark icon appears. Tap again to remove it

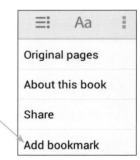

CHAPTER VII.

A MAD TEA-PARTY.

THERE was a table set out under a tree in front of the house, and the March Hare and the Hatter were having tea at it: a Dormouse was sitting between them, fast asleep, and the other

2 Bookmarks can also be added at any time by tapping on the **Menu** button and tapping on the **Add bookmark** button (once a bookmark has been added this action can be used to remove a bookmark too)

≡:   Aa   ⋮

Original pages

About this book

Share

Add bookmark

3 Bookmarks are included under the **Table of Contents** button.
Tap on the **Bookmarks** tab to view the available bookmarks

CHAPTERS	BOOKMARKS	NOTES

CHAPTER VII.A MAD TEA-PARTY.There was   95

4 Tap on a bookmark to go to that location

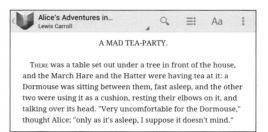

Alice's Adventures in…
Lewis Carroll

A MAD TEA-PARTY.

THERE was a table set out under a tree in front of the house, and the March Hare and the Hatter were having tea at it: a Dormouse was sitting between them, fast asleep, and the other two were using it as a cushion, resting their elbows on it, and talking over its head. "Very uncomfortable for the Dormouse," thought Alice; "only as it's asleep, I suppose it doesn't mind."

Definitions and Translations

Finding definitions

It is always satisfying to learn the meaning of new words, and when reading an ebook on your tablet this can be done at any point in the text:

1 Press and hold on the word for which you want the definition

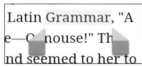

2 The definition appears at the bottom of the screen. Drag the box upwards to view it at the top of the page

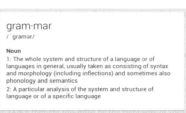

Translations

If you want to see what the text of a book looks like in a different language, this can be done with the **Translate** option:

1 Press and hold on a word and drag one of the markers to highlight a piece of text

2 Tap on the **Translate** button

3 Tap on this button to select a language

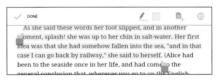

4 The translation is shown beneath the two selected languages. Tap on the **Done** button to return to the text

> **Beware**
>
> The translation option can be sightly hit or miss at times, so don't rely on it for any important translation tasks.

Adding Photos

Android tablets are great for storing and, more importantly, displaying your photos. The screen size of most tablets is ideal for looking at photos and you can quickly transform it into your own mobile photo album. In addition, it is also possible to share all of your photos in a variety of ways.

Obtaining photos

You can get photos onto your tablet in a number of ways:

- Most tablets have their own built-in camera which can be used to capture photos directly onto the device. The quality of these vary between makes of tablet: some are good quality cameras intended to be used for taking photos in a range of conditions. Others are mainly for use as a webcam for video calls (these are front-facing cameras)

- Transferring photos from your computer directly to your tablet, via a USB cable

- Transferring photos from your camera to your tablet. This is usually done by inserting your camera's memory card into a card reader connected to your computer and then transferring your photos

- Copying photos from an email

- Transferring photos from another device via Bluetooth

Beware

If you keep a lot of photos on your tablet this will start to take up its storage space.

136

Once you have captured or transferred photos to your tablet you can then view, edit and share them using the **Gallery** app. Photos in the Gallery app are stored in different albums, which are created automatically when photos are taken, transferred or downloaded from an email.

Gallery

Transferring photos

To transfer photos to your tablet directly from a computer or your camera's memory card:

1 Connect your tablet to the computer with the supplied USB cable

2 Access your photos on your computer, or insert the camera's memory card into a card reader and access this as an external device on your computer

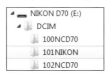

Don't forget

Photos copied directly from a camera's memory card will usually be in the JPEG file format, as will be the ones taken by the tablet's own camera.

137

3 Copy the photos that you want to transfer

4 Access your tablet in your computer's file structure

5 Navigate to the **Pictures** folder within **Internal Storage** and paste the copied photos here

Don't forget

Photos can also be copied straight from the camera if it is connected to the computer with a USB cable.

6 Open the **Gallery** app on your tablet to view the photos

...cont'd

Copying from email

Email is a good method of obtaining photos on your tablet: other people can send their photos to you in this way and you can also email your own photos, from a computer, a smartphone or another mobile device. To use photos from email:

1 Open the email containing the photo and tap on the **Attachment** tab

2 Tap on the **Load** button to view the photo in the email

3 Tap on the **Save** button to download the photo

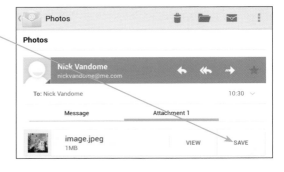

Don't forget

Once the **Download** folder has been created all other photos downloaded from emails will be placed here.

4 The photo will be saved in the **Download** folder within the Gallery app. This will be created if it is not already there

Viewing Albums

Once you have obtained photos on your Android tablet you can start viewing, managing and editing them.

1 Open the **Gallery** app. This displays the available albums. The contents of each album can be accessed by tapping on it

2 Tap on the **Menu** button at the top of the Album window to view the options

3 Tap on the **Select album** button to select one, or more, albums

Select album
Make available offline
Refresh
Settings
Help

4 Tap on an album to select it. Use these buttons on the top toolbar to, from left to right, share the album, share via Bluetooth or delete the album

Don't forget

The number of photos contained within albums is displayed in the right-hand bottom corner of each album.

Don't forget

Tap on the **Menu** button again in Step 4 to view the **Details** about the selected album. Tap on the **Done** button to complete the current action.

Viewing Photos

The photos in individual albums can be worked with and viewed in different ways:

1 Open an album. Tap here and select **Grid view** to view the photos in the album in a thumbnail grid

2 Select **Filmstrip view** as above to view the photos in the album as a filmstrip. Swipe left and right to view the photos

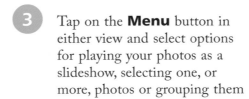

Hot tip

The **Group by** options in Step 3 are for grouping photos according to location, time taken, people and tagged photos. When you select one of these options you will be able to add the relevant information to a photo, so that it can then be used for grouping.

3 Tap on the **Menu** button in either view and select options for playing your photos as a slideshow, selecting one, or more, photos or grouping them

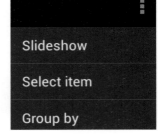

4 Tap on the **Select item** button in Step 3 on the previous page to select individual photos. Tap here and tap on **Select all** if you want to select all of the photos in an album

5 Tap on individual photos to select them. Tap on the **Menu** button to view the options for working with the selected image(s)

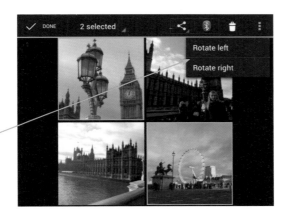

6 If only one photo is selected there is a greater range of Menu options than if multiple photos are selected

> **Hot tip**
>
> Select the **Set picture as** option in Step 6 to use the photo as your contact photo for your entry in the **People** app, or as your tablet's background wallpaper.

Editing Photos

Although the Gallery app is more for viewing photos, it does have a few editing options so that you can tweak and enhance your images. To access and use these:

1 Open an album, select a single photo and select **Edit photo** from the Menu button, or

2 Tap on a photo in an album to view it. Tap on this button to access the editing options

3 For both options above, the photo is displayed in **Edit** mode

4 Tap on this button to select a pre-set color option from the filmstrip below the photo

Don't forget

For the items in Steps 4 and 5, each time you select a new option the previous selection is discarded.

5 Tap on this button to select a frame option from the filmstrip below the photo

6 Tap on this button to access more editing options

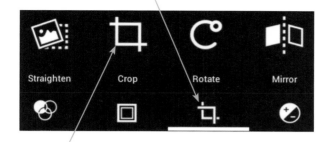

Straighten   Crop   Rotate   Mirror

7 To crop a photo, tap on the **Crop** button and drag the resizing handles as required

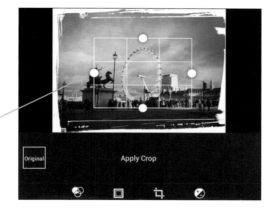

Original   Apply Crop

Hot tip

Most photos benefit from some cropping, to give the main subject more prominence.

143

8 Tap on this button to access the exposure options. Select an option and drag the slider to apply the effect

Apply Vignette 44

Don't forget

When you edit a photo and save it, a new image is created and the original remains intact.

9 Tap on the **Save** button to keep any changes that have been made

 SAVE

Sharing Photos

It can be great fun and very rewarding to share photos with friends and family. With an Android tablet this can be done in several ways:

 Select an album or open an individual photo and tap on this button on the top toolbar

 Select one of the sharing options. This will be dependant on the apps which you have on your tablet

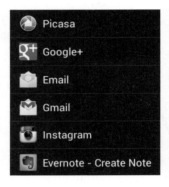

Beware

If you are sending photos by Bluetooth, the other device must have Bluetooth turned on and accept the request to download the photos when they are sent.

Sharing with Bluetooth

To share with another device with Bluetooth:

 Access an album or a photo and tap on the **Bluetooth** button on the top toolbar

 If your Bluetooth is not on, tap on the **Turn on** button to activate it

 Select the device with which you want to share your photo(s). These will be sent wirelessly via Bluetooth

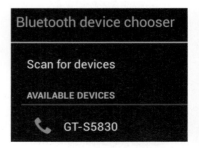

9 Keeping in Touch

Computers have transformed the way that we communicate and tablets have taken this to another level. Due to their portability, you can always have your communication options at your fingertips and this chapter shows how to make the most of your email, address book and calendars.

Email on Android

Email is now a standard feature in most people's digital world and it would be more of a talking point if there was not an email function on a mobile device rather than if there is. Android tablets cover several angles in terms of email, so that you can have access to any email accounts that you have. The two main options for email are:

- **Gmail.** This is the online webmail account provided by Google. When you create a Google Account you will also be provided with a Gmail account. This can be accessed directly from your Android tablet by tapping on the **Gmail** app. It can also be accessed from any device with Internet access

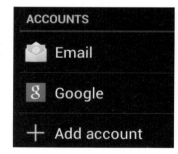

- **Using the Email app to add accounts.** This can be done if you already have an email account with another provider. This can be another webmail account or an IMAP or POP3 email account. New accounts can be accessed through the Email app and your Gmail account can also be viewed here

Details of email accounts are displayed under **Accounts** in the **Settings** app. Tap on the **Email** button to see details of all of the accounts that will be displayed by the **Email** app.

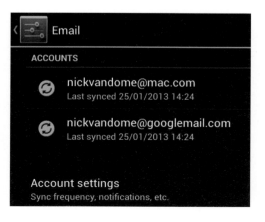

Don't forget

When you view details of email accounts you can also access the account settings for each one.

Adding Email Accounts

Email accounts can be added from the **Accounts** section of the **Settings** app, as shown in Chapter Three. In addition, they can also be added directly from the Email app. To do this:

1 Tap on the **Email** app

2 If no account has been added, the Email app will open at the **Account setup** page

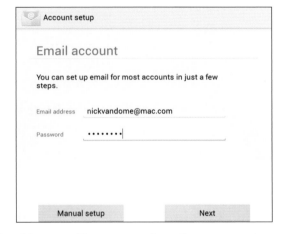

3 Enter the details of your email account and tap on the **Next** button. For some webmail accounts, the account should be added automatically and you will be taken to the screen in Step 6. This is the case if you are adding a Gmail account

4 If the account is not recognized automatically, tap on the **Manual setup** button in Step 3. Select the type of account that you want to set up and tap on the **Next** button

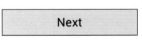

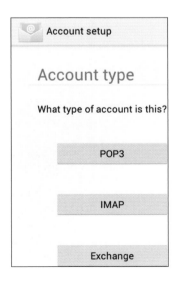

Hot tip

It is worth adding your Gmail account to the Email app if you already have an email account that you access from here. This way, all of your accounts can be accessed from one place.

147

...cont'd

5 For a manual setup, enter the details for the incoming and outgoing email servers. These can be obtained from your email provider. Tap on the **Next** button after each step

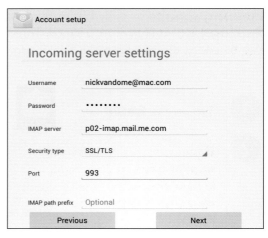

Don't forget

The **Account options** in Step 6 include making the selected account the default one for sending emails; specifying a notification for when email arrives; syncing email on the server; and automatically downloading email attachments over Wi-Fi.

6 After the server settings, select the **Account options** you want to use and tap on the **Next** button

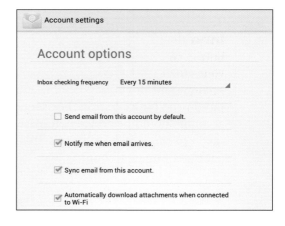

7 The setup is confirmed. Tap on the **Next** button to open the **Email** app and view your messages

Using Email

Viewing emails

Once you have added email accounts you can view them using the Email app.

1 Tap on the **Email** app

2 Tap here to view the available accounts. Tap on an account to view the messages in it, or tap on the **Combined view** button to view the emails in all of the accounts that have been set up

3 Tap on the **Inbox** button to view the messages in here

4 Tap on a message to view it. Use these buttons on the top toolbar to, from left to right, delete the email, move the email to another folder, view the next email and access the email's Menu settings

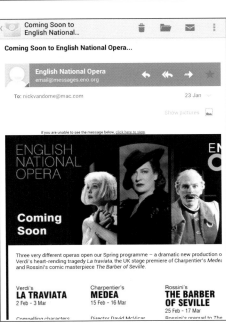

Don't forget

Unread messages are displayed on a white background; read messages appear on a gray background.

Don't forget

To select a specific email, check on the checkbox to its left and use the buttons on the top toolbar to organize it.

Hot tip

If you move to the next message by tapping on the envelope button on the top toolbar in Step 4, this marks the message from which you have just moved from as unread, i.e. it appears on a white background.

...cont'd

Sending an email

To send an email from the **Email** app:

1 Tap on the **New** button to create a new message

2 A new blank email is created

3 Enter a recipient, a subject and the body text of the email

4 Press and hold on a word and drag one of the blue markers to select text. Use the

top toolbar to **Cut** or **Copy** the text. Press and hold at another point to **Paste** the copied text

5 Tap on the **Send** button on the top toolbar to send the message. Tap on the **Save Draft** button to keep it and send it at a later date

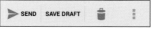

Email Settings

There are a number of settings that can be applied when using the Email app. They can be accessed from the Email Inbox or when an individual email is being composed.

1 In either the **Inbox** or an individual email, tap on the **Menu** button

2 Tap on the **Settings** button

3 Tap on an email account to view its settings

4 All of the settings for the selected account are displayed

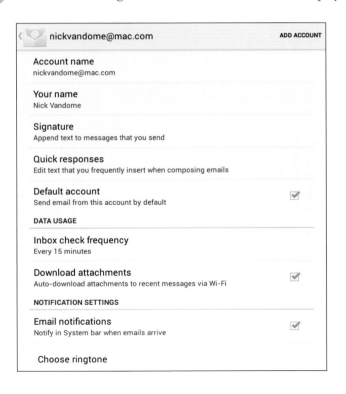

...cont'd

5 Tap on the **Account name** and **Your name** buttons to enter the respective names that will appear when recipients receive your emails

Account name
nickvandome@mac.com

Your name
Nick Vandome

6 Tap on the **Signature** button to add a message that appears at the end of your emails

Signature
Append text to messages that you send

7 Enter text for the signature and tap on the **OK** button to apply it

Signature

That's all from me

Cancel   OK

Don't forget

The **Quick responses** option is best used for unfamiliar words that may not be in the tablet's dictionary, such as real names.

8 Tap on the **Quick responses** button to enter specific words that you use regularly

Quick responses
Edit text that you frequently insert when composing emails

9 Tap on the **Create new** button

Create new

10 Enter the required word or phrase and tap on the **Save** button

Edit quick response

Vandome

Cancel   Save

11 When composing an email, tap on the **Menu** button and tap on the **Insert quick response** button to view the available options

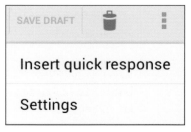

SAVE DRAFT

Insert quick response

Settings

12 Tap on a quick response word to insert it into your email

13 Check on the **Default account** checkbox to make this the default account for sending all emails

14 Check on the **Email notifications** checkbox to enable email notifications to appear on the Notifications Bar. Tap on the **Choose ringtone** button to select a sound for when a new email arrives

15 Tap on the **Inbox check frequency** button to specify a time period for how often the Email app checks for new emails from the server. Check on the **Download attachments** checkbox to use Wi-Fi to download attachments rather than 3G/4G

16 Tap on the **Server Settings** buttons to view and edit these, if required

SERVER SETTINGS

Incoming settings
Username, password and other incoming server settings

Outgoing settings
Username, password and other outgoing server settings

17 Tap on the **Remove account** button to delete the current email account

REMOVE ACCOUNT

Remove account

Hot tip

Set **Inbox check frequency** to a longer time period to reduce data usage. This could be useful if you have a 3G/4G connection and your data is delivered in this way instead of over a Wi-Fi connection.

153

Syncing Email Accounts

By default, email accounts are synced automatically. This means that they are synchronized with the server where the email accounts are stored. For instance, if you have a Gmail account it will automatically be synced on your tablet, so that if you send or receive emails when you are logged into your account elsewhere, any changes will also be applied on your tablet.

It is possible to turn off the automatic syncing for email accounts, if you want to perform this manually instead. To do this:

1 Tap on the **Settings** app

2 Under the **Accounts** section, tap on the **Email** button

3 Tap on one of the email accounts

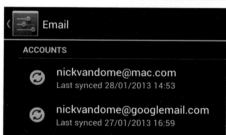

154

4 Check off the **Email** checkbox to turn off automatic syncing for the selected account

5 Tap on the **Menu** button and tap on the **Sync now** button to manually synchronize the account

Using Gmail

If you use Gmail as your default email account, you can access it directly with the Gmail app rather than using the Email app. This functions in a similar way to accounts through the Email app.

1 Tap on the **Gmail** app

2 The Gmail interface is similar to email accounts viewed in the Email app. Check on the checkbox next to an email to select it

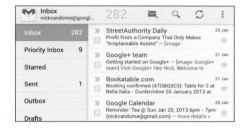

3 Tap the **Menu** button and tap on the **Settings** button to access the Gmail settings

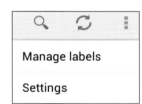

4 Tap on the **General settings** button or a specific account

5 The **General settings** have options for confirming certain operations such as deleting and sending, working with labels for emails, auto-fitting messages to the screen, clearing the search history and hiding pictures in messages, unless specified otherwise for certain senders

Don't forget

Tap on these buttons on the top toolbar to, from left to right, create a new message, search for an item in Gmail, refresh the current view and access the Gmail Menu:

Don't forget

The settings for a specific Gmail account include options for notifications, adding a signature and syncing settings.

Keeping an Address Book

An important part of staying in touch with people is having an up-to-date address book. On an Android tablet this can be done with the **People** app:

1 Tap on the **People** app

2 Your own basic details are already included from when you first set up your tablet

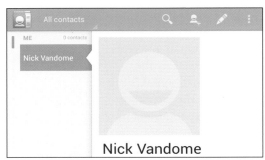

3 Tap on the **Edit** button to add more details to your own entry

4 Tap here and enter more information about your name and organization

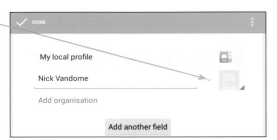

5 Tap on the **Add another field** button and select one of the options

Don't forget

Some tablets have their own default address book and calendar apps, which can be used in a similar way to the ones shown on the following pages. The People app and the Calendar app can be downloaded from the Play Store and these are the best ones to use if you want to sync your items with your Google Account.

Don't forget

Entries in the People app are also available online in your Google Account. This means that you can access them whenever you are online on another computer.

6 Tap here and take a photo with the tablet's camera or select one from the **Gallery** app. This will be your profile picture

7 Tap on the **Done** button

Adding contacts

To add more people to your address book in the **People** app:

1 Tap on the **New** button to add a new contact

2 Enter the details of the new contact

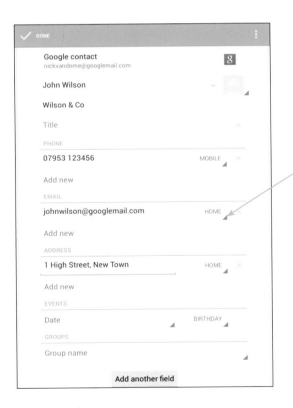

157

Hot tip

Tap on the arrow next to a field to access more options for that item.

3 Tap on the **Add another field** button if required

...cont'd

4 Tap here under the **Groups** heading to add the contact to a group

5 Check on the checkbox next to a group or tap on the **Create new group** button

Beware

Once a new group has been created you still have to add the contact to the group by tapping on the **Groups** heading as in Step 4 and then checking on the checkbox for the required group.

6 Enter a name for the new group and tap on the **OK** button

7 Tap on the **Done** button

8 The new entry is added under the **All contacts** headings

9 Tap here to add the contact to your **Favorites**

Working with Contacts

Once you have added a few contacts to the People app you can start organizing and managing them:

1 Tap here and tap on one of the options to view that selection according to **All contacts**, **Favorites** and **Groups**

2 If you select **Groups** in the step above, select the group that you want to view. Contacts that have been added to the group will be displayed

Don't forget

Tap on this button at Step 2 to create a new group and add contacts.

3 Tap on the **Menu** button to view the options for the People app

4 Tap on the **Share** button and select an appropriate app with which to share a contact. This is usually in the .vcf format, which is a text file containing the contact's details

Hot tip

Under **Display options** in **Settings**, tap on the **View contact names as** button to select whether a contact appears listed by their first name first, or their surname.

5 Tap on the **Settings** button in Step 3 to view the available **Display options** for your contacts

Using Your Calendar

The Calendar app can be used to add items such as events, meetings and birthdays. You can use these on your tablet and they will also be available online from any Internet-enabled computer or mobile device via your Google Account. To use the Calendar:

1 Tap on the **Calendar** app

2 The calendar is displayed and the view can be customized in a number of ways

3 Tap on this button to view the current day at any point

Don't forget

If you have Google Now activated on your tablet, your calendar events will be displayed here too.

4 Tap here at the top of the window to view the calendar by day, week or month or displaying all of the events that have been added (**Agenda**)

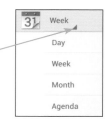

5 Swipe left and right on the main calendar to view different days, weeks and months. Swipe up and down on the month at the bottom of the window to see the next or previous month. Tap on a day to go to that point

Adding events

One of the main uses for the calendar is adding specific events:

1 In Day or Week view, tap and hold on a time slot and tap on the **New event** button, or

2 In any view, tap on this button to add a new event

3 The event will be linked to your Google Account email address

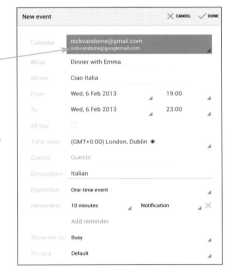

4 Enter the details for the event including: what, where, from, to and a description

5 Tap on the **Repetition** button if it is a recurring event and select the frequency

6 Tap here to select a time for a reminder for the event

7 Tap on the **Done** button to add the event to the calendar

Beware

The Calendar has to be linked to a Google Account in order for events to be added. If you do not have a Google Account, you will be prompted to add one before you can add an event.

161

Calling with Skype

Skype is a useful service that enables you to make voice and video calls to other Skype users, free of charge. To use Skype:

Beware

Skype voice and video calls are only free to other Skype users. Calls to non-Skype users have to be paid for.

Don't forget

Tap on the **Profile** button on the Skype Home screen to view and edit your own details.

1 Access the **Play Store** and download the Skype app

2 Open **Skype** by tapping on this button

3 If you already have a Skype account, enter your details, or tap on the **Create a Skype account** button to create a new account

4 Tap on the **Contacts** button to view your current contacts and add new ones

5 Tap on a contact to view their details and contact them by voice or video with these buttons at the top of the window, or by instant messaging with the text box at the bottom of the window

Adding Skype contacts

To add new contacts to call in Skype:

1 Access the **Contacts** section and tap on the **Search** button

2 Enter the name you want to search for and tap on the **Search** button again

3 Tap on the **Search Skype directory** button to search for the requested person

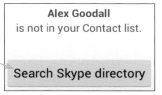

4 Matches for the requested person are shown. Tap on a person's name and tap on the **Add contact** button to send a contact request

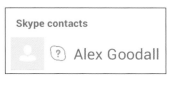

5 The selected person will then be sent a Skype request and they have to accept it before they become a full contact. Enter the text you want to use for the request, or use the default text that is pre-inserted

6 Tap on the **Send** button to send the message and request that the person accepts you as a contact on Skype

Don't forget

Once you add the Skype app to your tablet your Skype contacts will also be available in your **People** app.

Hot tip

Calls can also be made directly to numbers, using the keypad. Tap on this button on the main toolbar to access the Skype keypad.

Using Your Google Account

When you have a Google Account, the information that you enter into your communication apps such as Gmail, Calendar and People can also be viewed online whenever you are logged in to your account. To do this:

Don't forget

The sync options for the Gmail and Calendar apps are located within the **Settings** option. For the People app it is under **Accounts**.

Hot tip

To view the information from your People app through your online Google Account, click on the **Dashboard** link and click on the **Contacts** option.

1 Ensure that the sync options are checked on

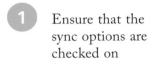

2 Log in to your Google Account online at **accounts.google.com/** and tap on the **Products** button

3 Tap on one of the relevant services to view the information from your Android app that has been synced with your Google Account, such as the Calendar

4 The details should be the same as for the item on your tablet

10 Browsing the Web

Web browsing on Android
tablets can be done with
different mobile browsers
This chapter looks at
obtaining browsers and also
some of the features found on
most mobile web browsers.

Android Web Browsers

Web browsing is an essential part of our digital world and on Android tablets this functionality is provided by a variety of web browsers customized for this purpose. They can usually display websites in two ways:

- Optimized for viewing on mobile devices, which are versions that are designed specifically for viewing in this format

- Full versions of websites, rather than the mobile versions, which are the same as used on a desktop computer

Different Android tablets have different default browsers but they all have the same general functionality:

- Viewing web pages

- Bookmarking pages

- Tabbed browsing, i.e. using tabs to view more than one web page within the same browser window

If you do not want to use the default browser that is provided with your tablet, there is a range of browsers that can be downloaded, for free, from the Play Store.

Enter **browsers for android** into the **Play Store Search box** to view the available options.

Don't forget

Mobile versions of a website usually have **m.** before the rest of the website address, e.g. **m.mynewssite.com**

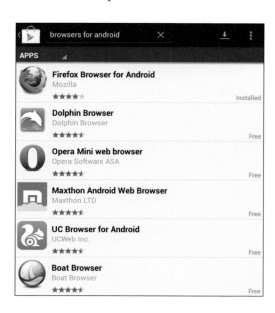

Opening Pages

Web pages can be opened on a tablet in an almost identical way as on a desktop computer or laptop. For some Android web browsers, there is a list of top sites when you open a browser or create a new tab. (The examples on the following pages are for the Chrome browser but other browsers operate in a similar way.)

1 When you first open Chrome there is a list of suggested sites and when you open a new tab there is a list of sites that you have already accessed. Tap on one to open it

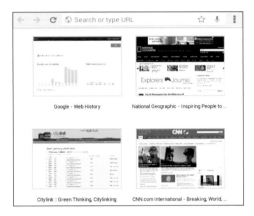

2 Enter a website address or search word into the **Search/Address** box. As you type, suggestions will appear. Tap on one to go to a list of results or tap on one to go to that website

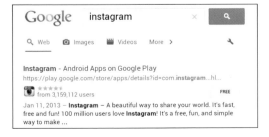

Don't forget

The Chrome browser can be downloaded from the Play Store if it is not already on your tablet. This is a Google product and integrates closely with other Google apps on your tablet.

Hot tip

Swipe outwards with your thumb and forefinger on a web page to zoom in on it; pinch inwards to zoom back out. Double-tap with one finger to zoom in and out too, but this zooms in to a lesser degree than swiping.

Bookmarking Pages

The favorite pages that you visit can be bookmarked so that you can find them quickly. To do this:

1 Open the page that you want to bookmark and tap on the star button in the **Search/Address** box

Hot tip

If bookmarks are saved into the **Mobile Bookmarks** folder they will be available on other mobile devices.

2 Tap in the **Name** box and enter a name for the bookmark

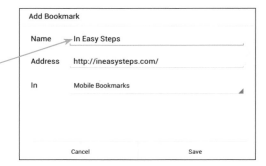

3 Tap in the **In** box to specify a folder into which you want to save the bookmark

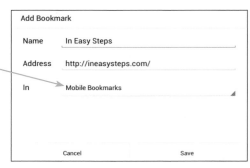

4 Check on a folder and tap on the **OK** button

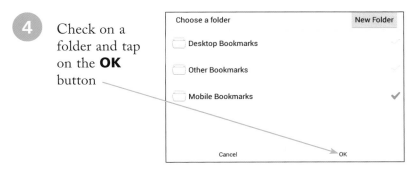

5 Tap on the **Save** button in the **Add Bookmark** window to save the bookmarked web page

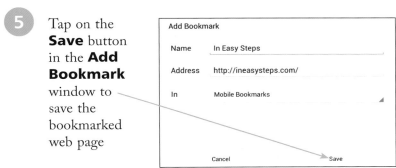

Don't forget

The **Menu** button can also be used to open a new tab. See page 171 for more details.

169

6 The star button turns dark which indicates the page is bookmarked

Viewing bookmarks

To view pages that have been bookmarked:

1 To view bookmarks, tap on the **Menu** button and tap on the **Bookmarks** button

2 The bookmarked pages are displayed in their relevant folders. Tap on one to open that page

Links and Images

Links and images are both essential items on websites; links provide the functionality for moving between pages and sites, while images provide the all-important graphical element. To work with these:

1 Tap and hold on a link to access its menu (tap once on a link to go directly to the linked page). The options on the menu are for opening the link in a new tab, opening it in a new

> http://edition.cnn.com/WORLD/
>
> Open in new tab
>
> Open in Incognito tab
>
> Copy link address
>
> Save link

tab that does not get recorded by the browser's History (**Open in Incognito tab**), copying it so that it could be emailed to someone, and saving the linked page as a download so that it can be viewed offline

2 Tap and hold on an image to access its menu. The options are for saving it (into the Gallery app), viewing it on its own on the current page (**Open image**) and opening

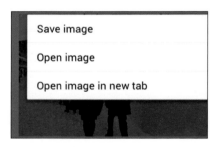

the image in a new tab. For both of the latter two options the image is displayed on its own on a page

Using Tabs

Tabs are now a common feature on web browsers and it is a function whereby you can open numerous pages within the same browser window. To do this:

1 Tap on this button at the top right-hand corner of the browser window to add a new tab, or

2 Tap on the **Menu** button and tap on the **New tab** button

New tab

3 Open a new page from the **Bookmarks** folders or by entering a web address or search word into the **Search/Address** box

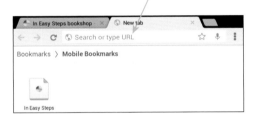

171

4 New tabs are opened at the top of the browser. Tap on the tab heading to move to that page

Hot tip

If there are a lot of tabs open, swipe left and right on the tabs bar to move between them all.

5 If too many tabs are opened for the available space on the screen, they are stacked on top of each other

6 Tap on the cross on a tab to close it

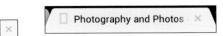

Being Incognito

If you do not want a record to be kept of the web pages that you have visited, most browsers have a function where you can view pages 'in private' so that the details are not stored by the browser. In Chrome this is performed with the Incognito function:

Don't forget

If the Incognito option is used, web pages will not be in the browser history or the search history.

Beware

If children are using your tablet then you may not know what they are looking at on the Web if they use the Incognito option. See Chapter 11 for details about restricting access to apps on your tablet.

1 Tap on the **Menu** button and tap on the **New incognito tab** button

2 The incognito page opens in a new tab, but any other open tabs are not visible (unless they are incognito too). Open a web page in the same way as for a standard tab

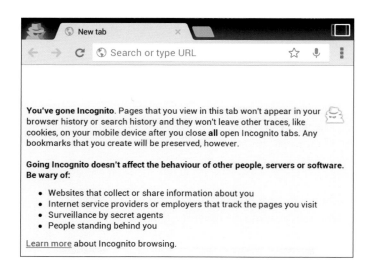

3 Incognito pages are denoted by this icon at the top left-hand corner of the browser

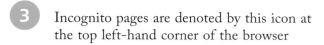

4 Tap on this button to toggle between incognito tabs and standard tabs. In each view, the tabs in the other view are not visible

Viewing Desktop Versions

A lot of web browsers for mobile devices have an option for viewing a mobile version of a website (if there is one) and in a lot of cases this is the default view. This enables the website to be viewed in the best format for the smaller screens used by tablets. However, even when the default view is for the mobile version it is still usually possible to view the desktop version, which is the one used by desktop computers and laptops. To do this:

1 In mobile view, items tend to be placed one above another. Scroll up and down to view the items

Don't forget

If there is no mobile version for a website, the default view will be the desktop one.

2 Tap on the **Menu** button and check on the **Request desktop site** checkbox to view the desktop version, if there is one available

Hot tip

If the **Request desktop site** checkbox is enabled this will apply to all websites that you subsequently visit, until you disable it.

3 If a desktop version is available it will have the full formatting for the website

Browser Settings

Mobile browsers have the usual range of settings, that can be accessed from the **Menu** button

1 Tap on the **Menu** button and tap on the **Settings** button

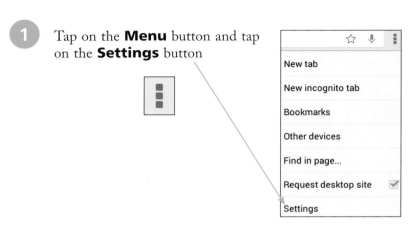

2 The full range of settings is displayed. Tap on each item to view its options. The **Advanced** settings are shown here

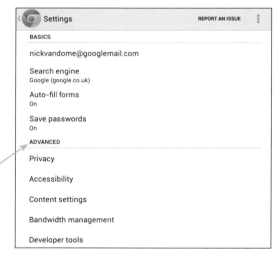

Don't forget

In the account settings you can select to view web pages that you have viewed on other devices and also sign in automatically to Google sites. At the top of the window you can also disconnect from your Google Account.

3 Tap on your account name at the top of the window to view details for syncing your account and using web pages viewed on another computer

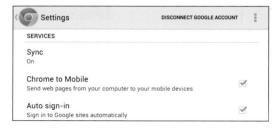

4 Tap on the **Search engine** button and tap on a search engine to set this as the default option for the browser

5 Tap on the **Auto-fill forms** button to make selections for how online forms are dealt with by the browser

6 Under the **Advanced** heading, tap on the **Privacy** button to specify how your browsing data is used

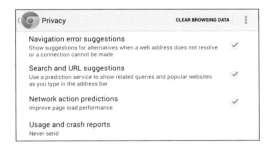

7 Under the **Advanced** heading, tap on the **Accessibility** button. Drag this button to increase the text size for viewing web pages in the browser

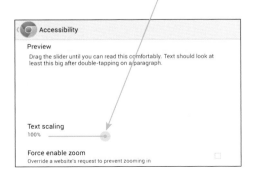

Beware

If other people are going to be using your account on your tablet do not turn on auto-fill options for credit or debit cards. If other people are using the tablet, it is best to set up individual accounts for them; see Chapter One for details.

Hot tip

In the **Privacy** settings you can remove your browsing data by tapping on the **Clear Browsing Data** button at the top of the window. You will then be able to select to clear your browsing history, cache, cookies, site data, saved passwords and auto-fill data.

...cont'd

A cookie is a small piece of data that is stored by the browser containing information about websites that have been visited. When you go back to the same website, the information is used from the cookie to identify your previous activity on the site. In most cases they improve the browsing experience. You'll also need to enable Javascript to get full functionality of most web sites.

If you use 3G or 4G services it is a good idea to set the bandwidth management to **Only on Wi-Fi** or **Never**, to avoid any possible data download charges.

8 Under the **Advanced** heading, tap on the **Content settings** button. Check on or off the options for cookies, Javascript and pop-up menus. Tap on the **Google location settings** button to specify whether Google apps can use your current location and the **Website settings** button to view settings for individual websites

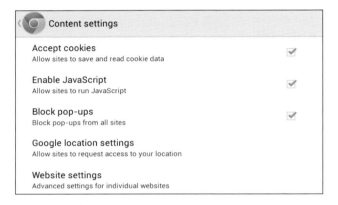

9 Under the **Advanced** heading, tap on the **Bandwidth management** button to specify how web pages are preloaded

10 For advanced users, who have experience of working with Android code, under the **Advanced** heading, tap on the **Developer tools** button to view options generally for debugging websites

11 Staying Secure

This chapter looks at security issues such as antivirus apps, dealing with a lost tablet and giving access to children.

Security Issues

Security is a significant issue for all forms of computing and this is no different for Android tablet users. Three of the main areas of concern are:

- **Getting viruses from apps.** Android apps can contain viruses like any other computer programs but there are antivirus apps that can be used to try and detect viruses. Unlike programs on computers or laptops with file management systems, apps on a tablet tend to be more self-contained and do not interact with the rest of the system. This means that if they do contain viruses it is less likely that they will infect the whole tablet

- **Losing your tablet or having it stolen.** If your tablet is lost or stolen you will want to try to get it back and also lock it remotely so that no-one else can gain access to your data and content. A lot of antivirus apps also contain a security function for lost or stolen devices

- **Restricting access for children.** If you have children who are using your tablet you will want to know what they are accessing and looking at. This is particularly important for the Web, social media sites, video sharing sites and messaging sites where there is the potential to interact with other people. There are no built-in parental controls on Android tablets, but there is a range of apps that can be used to provide children with access to just the apps that you want them to use. These can be locked so that children can only use the parentally-controlled apps and not gain access to the rest of the apps on the tablet

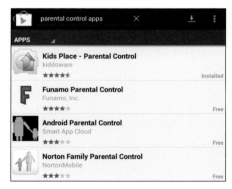

About Antivirus Apps

Android tablets are certainly not immune from viruses and malware and the FBI's Internet Crime Complaint Center (IC3) has even published advice and information about malicious software aimed at Android users. Some general precautions that can be taken to protect your tablet are:

- Use an antivirus app on your tablet. There are several of these and they can scan your tablet for any existing viruses and also check new apps and email attachments for potential problems

- Apps that are provided in the Play Store are checked for viruses before they are published, but if you are in any doubt about an app, check it online before you download it. If you do an online search for the app, any issues related it to should be available

- If you have a 3G or 4G connection on your tablet, turn this off when you are not using it. This will prevent any malicious software gaining access to your tablet through your 3G or 4G connection

- Do not download any email attachments if you are not sure of their authenticity. If you do not know the person who has sent the email then delete it

Functionality of antivirus apps

There are several antivirus apps available in the Play Store. Search for **android antivirus apps** (or similar) to view the apps. Most security apps have a similar range of features:

- **Scanning** for viruses and malicious software on your tablet

- **Online protection** against malicious software on websites

- **Anti-theft protection.** This can be used to lock your tablet so that people cannot gain unauthorized access, locate it through location services, wipe its contents if they are particularly sensitive and instruct it to let out an alert sound

- **Backing up and restoring.** Some information, such as your contacts, can be backed up and then restored to your tablet or another device

For some of the functions of antivirus and security apps a sign-in is required, such as for the anti-theft options.

Don't forget

A lot of antivirus and security apps are free, but there is usually a Pro or Premium version that has to be paid for.

Using Antivirus Apps

Antivirus apps operate by scanning your tablet for malicious software. There are also usually options for displaying an activity log and additional tasks that are available.

1. Open your antivirus app and tap on the protection or scan button

2. Details of the most recent scan are displayed. Tap on the **Scan** button to perform a new scan

3. Tap in the **Scheduled Scan** box to specify a time period for when scans are performed automatically

4. A countdown box shows the progress of the scan. This is where any results and actions are displayed

5 Tap on the app's menu (usually in the top right-hand corner) and tap on the **Activity Log** button, if there is one, to view tasks the app has performed

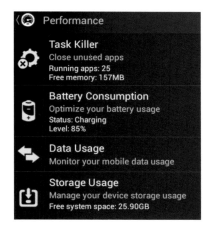

6 If there is a **Performance** option, tap on this to see the available items, including a **Task Killer**, that can be used to close down any unresponsive apps that are running

7 If there is a **Protection** option, tap on this to view additional scan options such as a **Deep File Scan** and **Safe Web Surfing**

Locating Your Tablet

The anti-theft function of antivirus and security apps can be used to locate your tablet if it is lost or stolen. This requires three elements to be in place:

● Location-based services have to be turned on, so that the anti-theft function can use this to locate it

● You have to sign in to the app for the anti-theft functionality. This requires a username (usually your Google Account email address) and a password. These details are used to log in to the associated website, see below

● You need to have access to the website associated with the app. This is where you will locate your tablet and perform any other tasks as required

To use the anti-theft function (this example is from the AVG AntiVirus app):

1 Open your antivirus app and tap on the **Anti-Theft** button

2 You will have to register with your Google Account details. Enter these and tap on the **OK** button

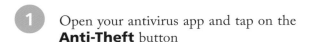

3 Check on the **Device Locator** checkbox to enable the app to use location-based services to locate your tablet

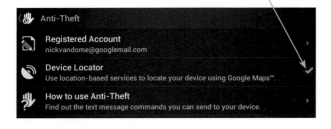

4 Some apps have options for sending text messages to your tablet, usually if it has a 3G/4G connection

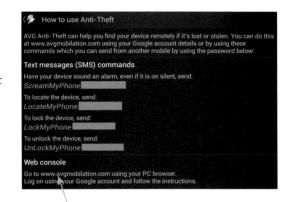

5 Open the app's associated website (this is usually detailed within the app on your tablet)

6 Log in to the website with your Google Account details

7 The website displays a map with your tablet's location. Click on the **Lock** button to remotely lock your tablet

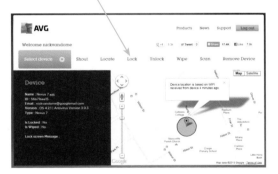

8 Enter a **Lock Device** password. This will be used to unlock your tablet when you get it back. Tap on the **Next** button to complete the process

Beware

Most anti-theft apps have an option for wiping the data from a lost or stolen device. Only do this if you are worried about someone getting access to the content on your tablet and if you know it has been backed up.

183

Parental Controls

Tablets are great devices for children: their portability and range of apps can make them the ideal device for playing games, watching videos and communicating with friends. However, these same benefits can also be a disadvantage as far as parents are concerned, as you may not always be able to see what your children are doing on the tablet as they may be in a different place from you.

There are no built-in parental controls on an Android tablet to restrict access to certain apps and types of content, but the Play Store contains apps that can be used for this. Some to look at are:

- Android Parental Control

- Best Parental Control Android

- Funamo Parental Control

- Kids Place

- Net Nanny for Android

- Norton Family Parental Control

Parental controls usually work in one of two ways:

- **Restricting apps.** When you open the app you can select items that you want restricted. This means that they can only be accessed with the entry of a password or a PIN code

- **Allowing apps.** When you open the app you can select items that you want made available to your child. These are displayed within the parental control app and this can only be exited with the use of a password or a PIN code

When you start using a parental controls app you will have to sign in with a password or PIN code and this is what is used to gain access to restricted apps or exit from a page that is displaying allowed apps.

Don't forget

There is a companion app to Kids Place that can be used to control access to online videos. It is called Kids Video Play.

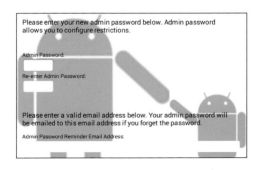

Please enter your new admin password below. Admin password allows you to configure restrictions.

Admin Password:

Re-enter Admin Password:

Please enter a valid email address below. Your admin password will be emailed to this email address if you forget the password.

Admin Password Reminder Email Address:

Restricting apps

For a parental control app that works by restricting apps (this example is for the Android Parental Control app):

1 Open your parental control app and check on the **Enable Restrictions** checkbox

2 Tap on the **Modify** button next to **Selected Apps**

3 Check on the checkboxes next to the apps that you want to restrict

Hot tip

Some parental control apps also have a safe browsing option that can be used to restrict certain types of content on websites.

185

4 The restricted apps are shown under the **Selected Apps** section. Tap on the **Modify** button to add or remove apps

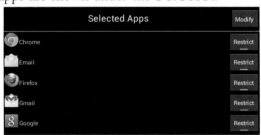

5 When someone tries to access a restricted app, the admin password has to be entered to gain access to the app

...cont'd

Allowing apps

For a parental controls app that works by allowing apps (this example is for the Kids Place app):

1 Open the parental controls app and tap on the **Select Apps** button

2 Enter the PIN that was created when you first started using the app and tap on the **OK** button

Hot tip

The tablet's home button can also be locked in some parental control apps so that the app cannot be exited in this way. This is usually accessed from within the app's settings options.

3 Check on the checkboxes next to the apps that you want to allow

Don't forget

When some apps are accessed through a parental control app the top Notifications Bar is visible. However, if anyone tries to access an item from it they will be informed that the action is not allowed and be directed back to the parental control app.

4 The restricted apps are shown on the app's Home Screen. They can be accessed here in the same way as normal. This screen can only be exited by entering the PIN in Step 2

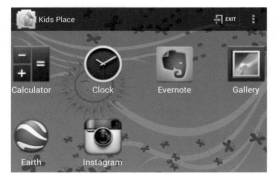

Index

H

I

J

K